THE USE OF DELFT3D TO SIMULATE THE DEPOSITION OF COHESIVE AND NON-COHESIVE SEDIMENTS IN IRRIGATION SYSTEMS

Shaimaa Abd Al-Amear Theol

Thesis committee

Promotor

Prof. Dr C.M.S. de Fraiture

Professor of Hydraulic Engineering for Land and Water Development

IHE Delft Institute for Water Education and Wageningen University & Research

Co-promotor

Dr F.X. Suryadi

Senior Lecturer in Land and Water Development
IHE Delft Institute for Water Education

Other members

Prof. Dr A.J.F. Hoitink, Wageningen University & Research

Prof. Dr N.C. van de Giesen, TU Delft

Prof. Dr J.A. Roelvink, IHE Delft

Dr A. Moerwanto, Special advisor Ministry Public Works, Jakarta Selatan, Indonesia

This research was conducted under the auspices of the SENSE Research School for Socio-Economic and Natural Sciences of the Environment

THE USE OF DELFT3D TO SIMULATE THE DEPOSITION OF COHESIVE AND NON-COHESIVE SEDIMENTS IN IRRIGATION SYSTEMS

Thesis

submitted in fulfilment of the requirements of

the Academic Board of Wageningen University and

the Academic Board of the IHE Delft Institute for Water Education

for the degree of doctor

to be defended in public

on Wednesday, 19 February 2020 at 3 p.m

in Delft, the Netherlands

by

Shaimaa Abd Al Amear Theol

Born in Baghdad, Iraq

Published by:

CRC Press/Balkema

Schipholweg 107C, 2316 XC, Leiden, the Netherlands

Pub.NL@taylorandfrancis.com

www.crcpress.com – www.taylorandfrancis.com

ISBN: 978-0-367-49691-3 (Taylor & Francis Group)
ISBN: 978-94-6395-231-6 (Wageningen University)

DOI: https://doi.org/10.18174/508011

ACKNOWLEDGMENTS

There are many people who were involved and gave lots of assistance and cooperation in this research. The author would like to deliver the gratefulness for their contribution.

Firstly, I would like to thank my Promoter and supervisor, Prof. Charlotte de Fraiture PhD, MSc who gave opportunity, ideas and support to this study, also I would like to thank Dr. Bert Jagers from Deltares for always sharing the ideas, knowledge and encouragement, the technical support and guiding me during this research and for his kind brotherly support. Also many thanks to my co-promotor Dr. F.X. Suryadi, PhD, MSc for sharing the ideas and guiding during my research.

I would like to thanks would like to thank the Iraqi Ministry of Higher Education and Scientific Research and the Ministry of Water Resources as well for funding my scholarship. Additionally, I would like to thank Deltares in Delft, the Netherlands for their technical support in providing the new version of the Delft3D and all modelling courses and workshops, it really appreciated especially Bert Jagers and Edward Melger.

I also express my gratitude for other staff of Hydraulic Engineering – Land and Water Development, I also express my gratitude for other staff of Hydraulic Engineering in Wageningen University and all guest lecturers who have taught me during my study in IHE-Delft.

Thanks to all my family of IHE who shared time together and did support each other and I would like to thank Prof. Bart Schultz, PhD, MSc for his help from the beginning, and my best friends Marielle Van Ervan, Mireia Lopez Royo, Zaki Shubber and Tonneke Morgenstond for their kind support. Many thanks to Professor Dano Roelvink, Roel Noorman, Loes Westerveen, Gordon de Wit, Jaap Kleijn and Lennard Teileman for their brotherly help. I would show their appreciation Dr. K.P. Paudel and SMIS for their field data and reports that have other data helpful for this study.

Above all, I would like to express my gratefulness for my husband Naser A. Kadhim and my son Ali, I dedicate this thesis as an insignificant gift for your endless love and sacrifices. Finally yet importantly, for others who were not mentioned but contributed to this thesis, I express my gratitude.

Summary

The deposition of sediments may threaten the performance and sustainability of irrigation systems by clogging canals and structures, disrupting water distribution, leading to unfair water distribution and high maintenance costs. Because of the high impact of sediment problems on irrigation performance and crop production, numerous studies have been conducted on how to deal with sedimentation in irrigation systems. Most of these studies concern non-cohesive (coarse) sediment, transported as bed load. These studies typically use 1D models. On the other hand, studies dealing with cohesive (fine) sediment are mostly done for rivers and estuaries; very few deal with irrigation systems. Cohesive sediment is generally transported in suspension and due to strong inter-particle forces and surface ionic charges, its behavior is more complex than non-cohesive sediment.

This research addresses two shortcomings of previous studies related to sediments in irrigation systems. Firstly, it uses a 2D and 3D model to simulate sediment deposition, where previous studies primarily used 1D models. The use of 2D and 3D models in irrigation systems is particularly important because of non-uniform flows around structures such as offtakes, weirs, and gates. This leads to asymmetric sedimentation patterns in cross-sections that are missed by 1D simulations. Secondly, this research simulates both cohesive, non-cohesive and a mix of cohesive and non-cohesive sediment, where previous studies mostly simulated pure cohesive or pure non-cohesive sediments. This is important for irrigation systems that draw water from natural rivers carrying a mix of both types of sediment.

The numerical model Delft3D was chosen for this purpose because it is well documented and proven reliable for the use in rivers and estuaries. It can be run in 2D and 3D mode and can simulate both cohesive and non-cohesive sediment. It can deal with networks and it can predict the morphological changes in the long term and has many other useful tools, such as Domain Decomposition, Flexible Mesh, and Real-Time Control.

After adapting the model Delft3D for the use in irrigation systems, the model was run for two canal systems in Sudan and Nepal. The findings showed the effect of the location of weirs and other structures; the impact of gate selection and operation on sediment deposition and erosion; and effect of the interaction of cohesive and non-cohesive sediment on sedimentation in irrigation canals. This knowledge is important in system maintenance and the development of gate operation plans that meet crop water requirements and at the same time minimizes sediment removal costs by alternating gates.

While Delft3D gave reasonable results, several challenges of the use of 2D and 3D models in irrigation canal systems were encountered. The running-time for complex networks is very long, even after using Domain Composition and Flexible Mesh. Furthermore, the model does not handle well the effect of sidewall friction and hence the model is not useful for small rectangular canals.

CONTENTS

1

Introduction

1.1 IMPORTANCE OF IRRIGATED AGRICULTURE

Irrigated agriculture plays an important role in global food production and some national economies. Especially in arid and semi-arid climates, irrigation is essential for successful crop cultivation. Because of the population increase, there is a large need to improve irrigation systems in order to meet the demand for food. Irrigation plays an important role in maintaining food supply for the growing population of the world, with around 270 million ha of irrigated land (i.e. 20% of the cultivated area) producing 40% of crop output (Paudel, 2010; Schultz & De Wrachien, 2002). However, about 14.5 million hectares of cultivated land per year are removed from agriculture due to urbanization, industrialization, waterlogging or salinity problems (Paudel, 2010; Schultz, 2002).

An irrigation scheme should not only be able to deliver the required amount of water to crops in the required time and water level. It should also be able to recover its operation and maintenance cost which is linked to the irrigation level of service. Maintenance costs can be high compared to the low-level ability of water users and farmers. Maintaining the quantity and quality of irrigation water and the service capacity of the existing irrigation systems is vital for crop production. To produce sufficient food for the increasing population and increase the productivity to assure future food security, it is essential to maintain irrigation water provisions to the canal command areas along with improving water management. To ensure sufficient water provision to meet crop water requirements and equitable water allocation for users (farmers), there is a great need for efficient operation and maintenance to improve the hydraulic performance of the canals and enhance the crop yields. Adequate water supply to crops can be achieved by improving water management through sediment management. This goal can be obtained if, among others, the effect of sediments in these canals can be reduced, where solving the sedimentation problems and/or reducing their negative impacts lead to improved efficiency of water allocation.

1.2 SEDIMENTS PROBLEMS IN IRRIGATION CANALS

Sediments can be classified into cohesive (fine) sediments, and non-cohesive (coarse) sediments. Cohesive sediments are composed primarily of clay-sized material and have strong inter-particle forces due to their surface ionic charge. Cohesive sediments are usually found in suspension mode. Non-cohesive sediments are composed of sandy material, which has weak interparticle forces. Non-cohesive sediments are usually found in the canal beds. Other differences between cohesive and non-cohesive sediments are listed in Table 1-1.

Table 1-1 Differences between cohesive and non-cohesive sediments

Cohesive (fine) sediments	Non-cohesive (coarse) sediments
Mostly in suspension. Encountered in the bed in very small quantities.	Transported in appreciable quantities as bed load. Their transport rate is the function of flow conditions (Simons & Fuat, 1992).
Sediment sizes smaller than 60μm (Hung et al., 2009).	Sediment sizes greater than 60 μm (Hung et al., 2009).
Flocs can be generated in the highly concentrated cohesive sediments, these flocs are hard to remove.	No flocculation.
In general, fine sediments have a flat plate or a needle shape and a high specific area (Partheniades, 2009).	Coarse sediments are almost rounded in shape.
In the case of cohesive sediments, the sediment transport predictors cannot be used, thus cannot determine the geometry of small canal "settling basins".	Standard sediment transport predictors are used to predict effects of changing canal slopes and cross sections when sediments are coarse (non-cohesive).
Cohesive sediments consolidate after deposition and require high rates of shear stress to re-suspended, which is not possible in small canals with low flow rate.	Non-cohesive sediments require low rates of shear stress before they again re-suspended.
Finest size fractions are transported through a settling basin.	Coarse sediments are trapped by the settling basin.
Cohesive sediments are relatively slow in settling for some hours (more than 12 hours) (Zac, 2012).	Non-cohesive sediments are faster in settling than cohesive sediments, they don't stay long in suspension they rapidly deposit in few minutes depending on the flow velocity (Zac, 2012).
Due to the ionic charge of cohesive sediments, they deposit faster in the salty water, while in fresh water, they stay in suspension longer or slowly deposited (Zac, 2012).	Non-cohesive sediments are not affected by the salty water (Zac, 2012).

Sedimentation in irrigation canals typically occurs in run-off-the river systems that are fed by rivers with high sediment loads. Sedimentation of canals can also occur from erosion of slopes next to the canal banks. The heavy particles of coarse sediment mainly settle in main and branch canals while fine sediments settle in smaller canals such as distributary canals or field watercourses. Sediment deposition highly affects the irrigation system performance and its sustainability. There are many reasons why sedimentation becomes a problem in irrigation systems, such as the lack of regular maintenance, the absence of optimal canal operation or, decreasing of flow discharge. Other reasons include evaporation from canals due to high temperature in semi-arid countries.

Sedimentation in irrigation canals cause many operational problems such as a reduction in conveyance capacity; blockage of the outlets and off-takes and disruption water distribution. Excessive sedimentation may raise the canal beds in the upstream part of the canal leading to higher water levels than the designed water level and lower water levels than designed in the downstream part of the canal. This will lead to the upstream outlets drawing more water than the quota while the downstream outlets get less water than the quota. In some cases raised bed and water levels lead to breached canal banks. In other cases, calibration of flow control structures and measuring devices is affected. All these cause problems of under- or oversupply, inequity, and, ultimately, a decline in the area that can be irrigated. This will adversely affect the production and farmers' satisfaction.

Unforeseen and unwanted sediment deposition and/or erosion in canals not only increase the operation and maintenance costs but also reduce the reliability of the services delivered. Solving sediment problems and getting rid of the unwanted erosion and deposition along the canal network requires substantial investments in money and labour. Sedimentation problems not only seriously affect the performance of the irrigation canals, but may also threaten their (financial) sustainability as well as reducing their productivity.

Sediment control approaches are initiated by selecting the proper diversion point and selecting the suitable structures at the river inlets to prevent unwanted sediment from entering the irrigation canals. The sediments that have already been entered the canals are then treated in different ways such as using coarse sediment traps, settling basins to get rid of them, or removal of sediment to a specific location where can be removed at a lower cost (Munir, 2011).Sedimentation in irrigation canals receives substantial scholarly attention due to the complex behaviour of sediments in canals. Many studies have been done to understand sediment behaviour in canals to develop approaches to reduce its impact on the canals. However, the vast majority of these studies deals with non-cohesive sediment while studies on cohesive sediments in irrigation canals are still limited.

1.3 OPERATION OF CANALS

In the design stage, the flow in irrigation canals is considered to be uniform and in equilibrium condition for the full supply of water, but this rarely happens in reality, bringing into question the validity of the assumptions made (Depeweg et al., 2015). Irrigation water demand is variable throughout the irrigation season as it depends upon the climatic conditions, soil moisture conditions, type of crops and the stage of crop growth. For this reason, irrigation canal networks carry variable amounts of water, often less than the design discharge. The design discharge, or canal capacity, can be defined as the maximum amount of flow that can be conveyed through canals, which depends on various factors like crop water requirement, irrigation methods, water distribution plans, flow control mechanism, and socio-economic settings. The change in the demand and the pressure for optimal water use every day leads to the need for proper canal operation. Canal operation consists of a package of organizational & economic and technical arrangements that ensure planned water distribution and full use of water resources for agricultural crops.

Therefore, the canal condition and the proper use of water during the canal operation should be considered (Renault et al., 2007). The operation practices include the following:

- Scheduling the efficient water in order to provide the required irrigation regime under specific meteorological conditions on certain land areas.
- Preventing excess water from flowing into the irrigation system and diverting it.
- Improving the system efficiency by controlling the water losses in canals.
- Organizing the accounting of irrigation water.
- Controlling the ideal water use and groundwater conditions.
- Controlling the crop management on irrigated lands.
- Liquidation of salinization and waterlogging on irrigated lands.

Flow control structures like gates, weirs, etc. are used to convey a certain amount of water or maintain a certain level of water for a specified period. These structures play a substantial role in sediment transport patterns either enhancing or reducing the deposition/erosion problems. Canal operation plays a significant role in the sedimentation processes since it affects the water level, velocity and flow along the canal, which in turn affects the sedimentation where changing the hydraulic regime affects sediment transport.

Water management becomes more difficult when there is sediment in irrigation systems (Mendez, 1998). Most irrigation management studies focus on non-cohesive sediment transport (such as sand). In case there is cohesive sediment (such as mud), the problem of management in irrigation canals becomes more complex.

Coarse sediments such as coarse sand and gravel can be excluded by using sediment control structures which are constructed at the head of runoff canals (Munir, 2011). However, these structures have little effect on sediment in suspension such as fine sand, silt, and mud because of the small size of these sediments. Hence finer sediment often is conveyed along the main channel and settled in lower levels such as distributary and/or field canals.

Settling basins are used in order to trap sediments and to make them deposited in certain locations where they can later be removed as a maintenance practice (Lawrence et al., 2001). A considerable amount of money is invested in order to remove the silting, however, in some schemes, sediment settles faster than they can be removed (Lawrence, 1998). The low settling velocities for sediments cause a long adaptation length before sediment concentration profiles adjust to a new set of hydraulic conditions after the disturbance and mixing introduced by a hydraulic structure as a gate (Lawrence, 1998).

Sediment transport rates depend on upstream and the local flow conditions. After deposition, the deposits of the cohesive sediments consolidate and they require high rates of shear stress before they are re-suspended (Lawrence, 1998). However, in small canals, where the bed shear is limited by small flows, it is difficult to re-suspend the consolidated sediments.

If the canal is not operated according to the design assumptions, the sediment problem cannot be solved, they can only be avoided or minimized if the operation and management plans are modified (Paudel, 2010).

The flow and sediment concentration highly affects the sedimentation and erosion processes. If some adjustments are made at a certain time intervals, a better scheme performance with less sedimentation in the upstream can be achieved. Additionally, the performance of the system with heavy sediment inflow can greatly be improved if the sediments are transported and deposited to the further area from where they can be removed with low cost (Jinchi et al., 1993).

1.4 MATHEMATICAL MODELS

There are several models that can simulate sediment transport. However, many of them are designed for rivers which render them unsuitable for particular features of irrigation systems, though some models can be adapted using user-written algorithms (Clemmens et al., 2005). Models used to simulate rivers cannot be directly applied to irrigation canals (Teisson, 1993).

Despite some functional and computational limitations in existing models, some have been modified for use in irrigation canals simulations. However, some critical limitations as width to depth ratio, and the roughness of the side slopes should be taken into consideration (Paudel, 2010).

Several factors with a significant impact on irrigation canals should be presented in the models that simulate sediment transport, such as the inflow of water and sediment. Several of these factors are not specified in river simulation models such as canal shape, existence of control structures as gates and weirs, and operation and maintenance practices. It is necessary to understand the interaction and influence of these factors in more than one direction to have a good understanding of sediment transport in irrigation systems.

Many sediment simulation studies for irrigation canals are using one-dimensional models that are relatively good from a hydrodynamic point of view, but not very accurate or representative regarding sedimentation in irrigation systems. Particularly in bends, near offtakes and around structures, flow patterns become 3-dimensional. This affects the sediment transport (both suspended load and bedload) causing spatial patterns in suspended sediment flows. 1D models can represent the sediment deposition or erosion in volume along the canals but cannot represent the sediment distribution in other directions.

1.5 PREVIOUS STUDIES & RESEARCH GAP

When sediments enter the canals some of them will be transported through the canal system to the fields and some will be deposited. In many irrigation schemes, excavators are used for sediment and aquatic weeds removal, but often there is a shortage in funds for maintenance to keep the system working properly. Sedimentation in irrigation systems has received substantial attention to understand sediment behaviour in canals and explore ways to reduce their negative impact. Researchers produced ideas and suggested methods to deal with non-cohesive sedimentation and to reduce the effect of it.

The hydrograph of water and sediment discharge has a great impact on the sediment degradation and aggregation processes in irrigation canals (Jinchi et al., 1993).

The irrigation clearance activities in Pakistan have been investigated by (Bhutta et al., 1996). They found that if they did the desilting campaign in the upper two-thirds of the canal, this will lead to significantly improve the hydraulic performance of the canals.

A new methodology has been developed by (Belaud & Baume, 2002) based on the use of a mathematical model. They illustrated this methodology for a secondary network in Sangro Distributaries System in South Pakistan, and proposed improvements in the design and desilting process in order to preserve the equity longer.

The design of the Sunsari Morang Irrigation System in Nepal and its impact on the sediments have been evaluated by (Depeweg & Paudel, 2003). They evaluate the effectiveness on sediment transport by using different operation plans and studied their effectiveness on sediment transport. Paudel (2010) proposed an improved rational approach for the design of alluvial canals which carrying sediment load, this approach can reduce the sediment deposition problem.

The net increase in bed level is defined as sedimentation, while the sedimentation rate is the deposition rate minus the erosion rate (Winterwerp & Van Kesteren, 2004).

The operation and maintenance become challenging in the scheme. The SETRIC model has been applied to simulate sediment transport in irrigation canals in Nepal and Indonesia by (Sherpa, 2005) and (Sutama, 2010) respectively. They evaluated this model by using different operation and sediment input in irrigation canals.

A mathematical model has been developed and applied to simulate the sediment in irrigation canals by (Jian, 2008), where the adopted model can be used to predict the non-uniform sediment movement in irrigation canals.

The impact of the operation on the sediment deposition in the USC-PHLC Irrigation System in Pakistan was studied by (Munir, 2011). He found that the sediment deposits during low crop water requirement periods can be re-entrained during peak water requirement periods and he suggested an improvement in the canal operation.

Cohesive sediments transport in irrigation canals under different operation plan has been tested by (Osman, 2015) through using a one-dimensional model developed by her based on the sub-critical, quasi-steady flow in which can simulate sediment transport under non-equilibrium conditions. The best option of operation is to apply the continuous operation system, which can reduce the deposition by 55% when compared to the night storage system (Osman, 2015).

However, the majority of these studies dealt with non-cohesive sediment behaviour, but few studies have been done for cohesive sediments only and almost none have been done regarding the mixed sediments in irrigation systems.

Cohesive sediments affect the management of water; the physical processes of the cohesive sediment transport are still not well understood. Additionally, the combination between the hydrodynamic, cohesive sediment properties and biological processes makes the prediction of cohesive sediment dynamics complicated. Many parameters have an influence on the dynamics of cohesive sediment. However, these parameters cannot be specified theoretically. For this reason, cohesive dynamics are solved empirically and hence, the dynamics of cohesive sediment still not clear yet (Lopes et al., 2006).

The incomplete knowledge of fundamental processes such as deposition, erosion, and consolidation of cohesive sediment, leads to proportional failure in obtaining the quantitative results, not because of the well experienced numerical techniques. Many researchers used 1D-models in their study, but 1D-models may not be representative regarding the sediment behaviour, location of the accumulation and sediment patterns especially within the cross-section and near hydraulic structures. In general, due to the complex physical processes of cohesive sediments, there is a lack of knowledge and a great need to do more studies using 2D/3D models in order to reinforce the understanding of cohesive sediment behaviour especially under different operation conditions.

The majority of numerical sediment simulation models are dealing with non-cohesive sediment in rivers, coastal and irrigation systems. Models dealing with cohesive sediments are suitable and applied for rivers and estuaries, but not in irrigation canals. There are some similarities between rivers and irrigation canals, but at the same time still, there are major differences between them (Table 1-2).

Table 1-2 Differences between rivers and irrigation canals

Rivers	Irrigation canals
b/h ratio is rather big	b/h ratio is smaller
Wave have effect in the river	There is no effect of wave
There are no control structures in the rivers	Presence of many control structures to control water level and discharge
Different directions for flow	Unidirectional flow
Mathematical models which are used: 1D, 2D and 3D	Mathematical models which are used: 1D
There is big capacity and velocity	Limited capacity and flow velocity
There is no side banks, and no affect on velocity distribution	Great influence of side banks on the velocity distribution

These differences make mathematical models which are used in rivers not comfortable to be used in irrigation systems, and not because of limitation in these models which are well designed. Lastly, a mathematical model that deals with cohesive sediments in the irrigation system was developed by (Osman, 2015), however, this model is limited by being one-dimensional model.

From these shortages, the gaps are:

❖ Mathematical models, for better insights and well understood, we need to model cohesive sediments in irrigation canals in 2D perspective. If there is an existing model which is developed primarily for rivers, is it possible to be used in irrigation systems simulations? If yes, how can we adapt it for the use in irrigation canals?

❖ The canal operation effects on cohesive sediments, how can we through finding suitable operation, control and reduce the cohesive sedimentation. And what this suitable operation which will reduce the negative impact of cohesive sediments is, as well as providing efficient water delivery enhancing crop production with reducing maintenance costs.

1.6   RESEARCH OBJECTIVE

1.6.1 Main and specific objectives of the study

The main objective is using a 2D/3D simulation model to study the impact of canal operation on cohesive and non-cohesive sedimentation to support optimal canal operation which can reduce the negative effects of sediments.

The specific objectives are to:

❖ Use suitable 2D/3D mathematical model for irrigation canals.
❖ Analyze the cohesive sediment transport process under actual irrigation canals conditions.
❖ Analyze the existing canal operation in order to find the relationship between water and cohesive sediment transport and water management in the canal.
❖ Evaluate different structures effect on the sediments' distribution and transport.
❖ Evaluate various canal operation scheme and to recommend possible improvements and canal operation plan for better water and sediment management.

1.6.2 Research questions to be identified

To deal with the sedimentation problems in an irrigation system, the following research questions are raised:

1. Since there are some similarities between rivers and irrigation canals, the questions raised are:

   A- Can a 2D/3D model which is already producing adequate results for sediment simulations in rivers be used in irrigation systems?

B- How can this model be adapted for simulating cohesive, non-cohesive and mixed sediment transport in irrigation canals? (Chapter 2).

2. Based on the known differences in shape, size between the cohesive and non-cohesive sediments, the questions raised are:

A- How will cohesive sediments differ from non-cohesive sediments and their mixture regarding their distribution, canal bed morphology development, their sensitivity, and deposition and erosion in different locations?

B- What is the effect of the interaction between cohesive and non-cohesive sediment? (Chapter 3).

3. Regarding the canals operation in irrigation systems, the questions raised are:

A- What is the effect of gate selection and gate operation on the non-cohesive sediment transport? (Chapter 4).

B- Considering existing structures in irrigation systems to control the water level and amount of water to be diverted to the branch canals, the question is: What is the effect of different structures (weirs and gates), and what is the effect of gate operation on the cohesive sediment transport? (Chapter 5).

1.7 METHODS

Modelling using Delft3D

There several models that can simulate sediment transport. In this research Delft3D has been chosen because of its multiple advantages. Delft3D is a multi-dimensional (2D and 3D) developed by Deltares (Deltares, 2016). It has many modules of which, the Delft3D-FLOW, can calculate steady and non-steady flow and transport phenomena in 2D and/or 3D approach with the existence of weirs, and gate operation for both cohesive and non-cohesive sediment transport.

Governing equations in Delft3D

In the design phase, when schematizing the water flows in irrigation canals, two important considerations should be made. The first concerns the hydraulics and operational aspects. Due to the changes in water requirements and the gate operations to satisfy water demand and maintain the desired water levels, the water flows become non-uniform. The second consideration concerns sediment transport, since the changes in the morphology of sediments are slower than changes in water flow in time and space (Depeweg & Méndez, 2007).

For the hydraulic aspects, the Reynolds averaged Navier Stokes equations are solved by Delft3D-FLOW, which calculates non-steady and steady flow and provides the hydrodynamic basis for morphological computations. For the sediment aspect, the bedload and suspended load transport of non-cohesive sediments and the suspended load of cohesive sediments are supported by the sediment transport and morphology module (Delft3D-MOR). To schematize between kinds of sediments, 'mud' is recognized as cohesive suspended load transport, while 'sand' is recognized as non-cohesive bedload and suspended load (Luijendijk, 2001).

The transport of suspended sediment is calculated by solving the 3D advection-diffusion (mass-balance) equation for suspended sediment

$$\frac{\partial c^l}{\partial t} + \frac{\partial u_c{}^l}{\partial x} + \frac{\partial v_c{}^l}{\partial y} + \frac{\partial (w - w_s{}^l) c^l}{\partial z} - \frac{\partial}{\partial x}\left(\varepsilon^l{}_{s,x}\frac{\partial c^l}{\partial x}\right) - \frac{\partial}{\partial y}\left(\varepsilon^l{}_{s,y}\frac{\partial c^l}{\partial y}\right) - \frac{\partial}{\partial z}\left(\varepsilon^l{}_{s,z}\frac{\partial c^l}{\partial z}\right) = 0 \quad \textit{1-1}$$

Where:

$c^{(l)}$ = mass concentration of sediment fraction (L) (kg/m^3)

u, v and w = flow velocity components (m/s)

$\varepsilon_{s,x}{}^{(l)}$, $\varepsilon_{s,y}{}^{(l)}$ and $\varepsilon_{s,z}{}^{(l)}$ = eddy diffusivities of sediment fraction (L) (m^2/s)

$w_s{}^{(l)}$ = hindered velocity

But in irrigation canals there are no eddies; therefore the last three terms will be omitted from Equation (1.1) and it will become

$$\frac{\partial c^l}{\partial t} + \frac{\partial u_c{}^l}{\partial x} + \frac{\partial v_c{}^l}{\partial y} + \frac{\partial (w - w_s{}^l) c^l}{\partial z} = 0 \qquad \textit{1-2}$$

Delft3D-flow uses Partheniades-Krone formulation (Partheniades, 1965) to calculate the fluxes between the water phase and the bed for deposition and erosion of cohesive sediment fractions (Deltares, 2016).

Erosion formula:

$$E^l = M^l * S(\tau_{cw}, \tau_{cr,e}^l) \qquad \textit{1-3}$$

Where:

E^l : Erosion flux [kg m^{-2}s^{-1}];

M^l: User-defined erosion parameter [kg m^{-2}s^{-1}];

$S(\tau_{cw}, \tau_{cr,e}^l)$: Erosion step function [-];

τ_{cw}: Maximum bed shear stress [N/m^2];

$\tau_{cr,e}^l$: User-defined critical erosion shear stress [N/m^2].

$$S(\tau_{cw}, \tau_{cr,e}^l) = \begin{cases} \left(\frac{\tau_{cw}}{\tau_{cr,e}^l} - 1\right) & \textit{when} \quad \tau_{cw} > \tau_{cr,e}^l \\ 0 & \textit{when} \quad \tau_{cw} < \tau_{cr,e}^l \end{cases} \qquad \textit{1-4}$$

Deposition formula:

$$D^l = W_s^l * C_b^l * S(\tau_{cw}, \tau_{cr,d}^l) \qquad \textit{1-5}$$

$$C_b^l = C^l(Z = \frac{\Delta Z_b}{2}, t) \qquad \textit{1-6}$$

Where:

C_b^l: Average sediment concentration in the near bottom computational layer [kg/m^3]

D^l : Deposition flux [kg m^{-2}s^{-1}]

$S(\tau_{cw}, \tau_{cr,d}^l)$ Deposition step function, τ_{cw} Maximum bed shear stress [N/m^2]

$\tau_{cr,d:}^l$ User-defined critical deposition shear stress [N/m^2]

W_s^l : Fall velocity (hindered) [m/s]

Z_b: Depth down to the bed from a reference height [m]

$\Delta Z_{b:}$ Thickness of the bed layer [m]

$$S(\tau_{cw}, \tau_{cr,d}^l) = \begin{cases} (1 - \frac{\tau_{cw}}{\tau_{cr,d}^l}) & \textit{when} & \tau_{cw} < \tau_{cr,d}^l & \textit{1-7} \\ 0 & \textit{when} & \tau_{cw} > \tau_{cr,d}^l \end{cases}$$

Although the model equations allow for the specification of a critical shear stress for deposition, various researchers (Chan et al., 2006; Sanford & Halka, 1993; Winterwerp & Van Kesteren, 2004) indicate that it does not exist in nature. The authors therefore used default value of $t_{cr, d}$ = 1000 N/m^2. The high value for $t_{cr, d}$ causes $S(\tau_{cw}, \tau_{cr,e}^l)$ to be effectively equal to 1, therefore we can neglect this term from eq.4 and the equation will be as below:

$$D^l = W_s^l * C_b^l \qquad \textit{1-8}$$

The bedload and suspended load transport of non-cohesive sediments are supported by the sediment transport and morphology module. For the computation of the non-cohesive sediment behaviour the Delft3D model applies the approach developed by Van Rijn (1993). Van Rijn predicts the sediment transport as bedload and suspended load, he used a reference height (a) to differentiate between these loads. Sediment transport below this reference height is treated as bedload transport and above it as suspended load transport. The reference layer which is called kmx-layer is above the Van Rijn reference height, where the sediments in this layer transfer between bed and the flow. In the layer(s) that lie below the kmx-layer, the sediment concentrations are presumed to be adjusted to similar concentrations of the reference layer. The quantity of sediment entering the flow due to upward diffusion from the reference level, as well as the quantity of sediment dropping out of the flow due to sediment settling are modelled each half time-step by the source and sink terms.

The advection-diffusion equation solves the sink term implicitly, whereas the source term is solved explicitly. In order to determine the sink and source terms of the kmx-layer, the concentration and concentration gradient at the bottom of this layer needs to be approximated, and the Rouse profile is assumed to be standard between the reference height (a) and the center of the kmx-layer. Sink and source terms are calculated as follows:

$$C^l = C_a{}^l \left[\frac{a(h-z)}{z(h-a)}\right]^{A^l} \qquad \textbf{\textit{1-9}}$$

In the Delft3D model, the reference height can be represented as:

$$a = min \, [max \, \{f * ks, \, 0.01h\}, \, 0.2h] \qquad \textbf{\textit{1-10}}$$

Where;

$A^{(l)}$ = Rouse number;

a = Van Rijn's reference height [m];
$c^{(l)}$ = concentration of sediment fraction (l) [kg/m^3];
ca $^{(l)}$ = reference concentration of sediment fraction (l) [kg/m^3];

f = user define proportionality factor [-];
h = water depth [m];

K_S= roughness height [m].

Erosion formula:

$$E^{(l)} = \frac{\propto_2{}^l \varepsilon_s{}^l C_a^l}{\Delta z} - \frac{\propto_2{}^l \varepsilon_s{}^l C_{kmx}^l}{\Delta z} \qquad \textbf{\textit{1-11}}$$

Where the first term is $(source\,{}^l_{erosion})$ and the second term is $(sink\,{}^l_{erosion})$

Deposition formula:

$$D = \propto_1{}^l C_{kmx}^l w_s{}^l \qquad \textbf{\textit{1-12}}$$

$$(source\,{}^l_{deposition}) = \propto_2{}^l C_a^l\left(\frac{\varepsilon_s{}^l}{\Delta z}\right) \qquad \textbf{\textit{1-13}}$$

$$(sink\,{}^l_{deposition}) = \left[\propto_2{}^l \left(\frac{\varepsilon_s{}^l}{\Delta z}\right) + \propto_1{}^l w_s{}^l\right] C_{kmx}^l \qquad \textbf{\textit{1-14}}$$

Where:

C_a^l : Reference concentration of sediment fraction (l) [kg/m^3];

C_{kmx}^l Average concentration of the kmx cell of sediment fraction (l) [kg/m^3];

$w_s{}^l$: Settling velocity [m/s];

Δz: Difference in elevation between the center of the kmx cell, Van Rijn's reference height: (Δz = zkmx − a) [m];

$\propto_1^1$: First correction factor for sediment concentration [-];

$\propto_2^1$: Second correction factor for sediment concentration [-];

ϵ_s^1 : Sediment diffusion coefficient evaluated at the bottom of the kmx cell of sediment fraction (l) [-].

1.8 STRUCTURES OF THE THESIS & SCOPE OF THE STUDY

The Delft3D model will be adapted for the use in irrigation systems simulations and will be used to predict the morphological developments of canals bed under different operation scenarios to evaluate the operation impacts on the cohesive sedimentation and to propose recommendations to the designers of new canals to take into account the sedimentation impacts, and to propose the optimal operation plan that ensures less deposition and relatively good canal performance.

By using 2D/3D-modelling, this research will focus on cohesive, non-cohesive and mixed sediments and their behaviour, how they affect the irrigation systems and how they are transported along the canals and within the cross-sections. The effect of the sediment accumulation along the canals on the stability of the water level will be described. The difference between the cohesive and non-cohesive sediment transport will be clearly explained. The effect of the gate selection and operation on the sedimentation processes will be well studied in this research. Also, it will address the impact of structures on sediment transport.

Chapter 1 gives an overview of the research as general introduction, problems statement, research approach and objectives, the scope of the study and the structure of the thesis.

Chapter 2 tests the suitability of Delft3D model for the use in irrigation systems simulations.

Chapter 3 presents the application of Delft3D to show the differences between cohesive and the non-cohesive sediment behaviour in irrigation systems and their interaction.

Chapter 4 presents the effect of using different gate operation plans on the non-cohesive sediments in the irrigation system with the existence of settling basin (case study: SMIS Scheme in Nepal).

Chapter 5 displays the impacts of different structures and using different operation plans on the cohesive sediments deposition in irrigation systems (case study: Gezira Scheme in Sudan).

Chapter 6 presents the conclusion, reflection, research contribution and recommendations for further studies.

2

THE USE OF DELFT3D FOR IRRIGATION SYSTEMS SIMULATIONS

This chapter is based on:

Theol, A, S., & Jagers, B., & Suryadi, F., and de Fraiture, C. (2019). The use of Delft3D for Irrigation Systems Simulations. Irrigation and Drainage, 68(2), 318-331. doi:10.1002/ird.2311

ABSTRACT

Irrigation systems performance and sustainability are affected by sediment deposition. Cohesive sediment (suspended load) is an important problem in irrigation canals and its behaviour is significantly different from that of non-cohesive sediment (bed load). Most studies on sedimentation in irrigation systems deal with non-cohesive sediment. Studies on cohesive sediments are mostly done in rivers and estuaries, but not in irrigation canals. The few existing studies on cohesive sediment in irrigation canals are limited by their use of 1D models. Therefore, in this chapter we test whether an existing 3D model that was designed for rivers and estuaries can be used in irrigation canals. Delft3D was identified as a suitable model. Simulations were done for different sizes and configurations of the irrigation network. After some adaptations to the model, the simulations of different scenarios provided promising results. From a hydrodynamic and morphological point of view the Delft3D model was able to realistically represent water and sediment flows in a hypothetical canal set-up, consisting of a main canal, a branch canal and several hydraulic structures. Some challenges remain in the use of Delft3D for irrigation canals, in particular regarding wall roughness in small rectangular canals and computation times for complex systems. However, these challenges are not insurmountable and the advantages of using Delft3D are clearly shown in this chapter.

2.1 INTRODUCTION

Sediments can cause serious operational problems in irrigation systems. Raised canal bed levels may lead to raised water levels upstream of the canal so that fields upstream get more water than their quota and downstream fields get less. Thus, sedimentation in canals can cause problems of undersupply, unfairness and an inevitable decline in the irrigated area, affecting production and farmers' satisfaction (Munir, 2011; Paudel, 2010). Existing control structures in irrigation systems affect hydraulic parameters such as velocity and bed shear stress, and hence have a big impact on the rate of deposition or erosion of sediments.

While there have been numerous studies and simulations concerning sediments in irrigation canals, these mainly dealt with non-cohesive sediment or bed load (see e.g. Paudel, 2010 and Munir, 2011). Studies on cohesive sediment (i.e. sediment in suspension) in irrigation canals are few. Osman et al. (2016) used a case study in the Gezira irrigation scheme in Sudan. After they concluded that none of the existing models were suitable for cohesive sediment in canals, they developed the Fine Sediment Transport (FSEDT) model. Testing different scenarios of canal operation using FSEDT, they formulated strategies of water management to reduce deposition in irrigation canals (Osman, 2012); Osman et al. (2016). Belaud and Baume (2002) applied the Simulation of Irrigation Canals (SIC) model to simulate cohesive sediment in a secondary network of the Sangro Distributary System in Pakistan. The canal was equipped with sensors, actuators and a SCADA system (Supervisory Control And Data Acquisition). They recommended improvements in the design and desilting process in order to maintain equity for a longer period of time.

The studies by Belaud and Baume (2002) and Osman and Schultz (2016) employed 1D models. To simulate hydrodynamic flow in canals 1D models are suitable tools, but in the case of simulating fine sediments in irrigation systems 1D models may not be representative. Sediments in suspension do not move in one direction with the flow. Particularly at bends, near offtakes and around structures suspended sediment flows in different directions and settles in different parts of the canal cross section. This chapter hypothesizes that for simulating the behaviour of cohesive sediment in canals, 2D and 3D models are more suitable.

The processes governing the behaviour of cohesive and non-cohesive sediment differ significantly. Most research on cohesive sediments has been undertaken in rivers (Krishnappan, 2000) and estuaries (Van der Wegen et al., 2011). Gebrehiwot et al. (2015) evaluated existing flood and sediment management practices in the Aba'ala spate irrigation system. Using the Delft 3D model, they identified alternative intake designs and locations for optimum water and minimum sediment intake.

There are similarities between rivers and irrigation canals such as the bed shear stress and the friction forces which are the dominant factors in the flow. There are also important differences such as the b/h ratio and the side slope (Mendez, 1998). Other differences are: the presence of flow control structures in irrigation canals due to the need to control level and discharge, and the considerable influence of side walls on velocity distribution (Depeweg & Méndez, 2007).

This means that results from the simulation of cohesive sediment behaviour in rivers are directly transferable to irrigation settings.

2.1.1 Delft3D

Because no 2D or 3D models exist for simulating flows and cohesive sediments in irrigation canals, this chapter uses a 3D model originally designed for rivers, and adapts it for irrigation systems. There are several 2D and 3D models that simulate cohesive and non-cohesive sediments in rivers such as the SED2D WES model, which is a finite-element model developed by the US Army Waterways Experimental Station.

This model can simulate cohesive and non-cohesive sediments in rivers, but it considers a single effective grain size in each simulation and is not freely available. The Mike 21C model, which is an integrated river morphology tool developed by the Danish Hydraulic institute (DHI) in 2009, is designed and used for non-cohesive sediments but not for cohesive sediments. Some researchers used the Delft3D model in their research, i.e. for rivers (De Jong, 2005; Flokstra et al., 2003; Kemp, 2010) and for estuaries (Lesser, 2009; Van der Wegen et al., 2011). The Delft3D model presents some disadvantages, such as the effect of wave asymmetry on bed load transport, and wave forcing and a roller model varying the timescale of wave groups (Luijendijk, 2001). However, these disadvantages are not relevant to irrigation canals.

In this chapter the Delft3D model is chosen to simulate cohesive sediment in irrigation canals because it is freely available, well documented and tested, and simulates hydrodynamic flow for rivers and computes sediment transport for (cohesive and non-cohesive) sediments. In addition, it can deal with networks and the existence of structures and it predict the long term for morphological changes in beds.

The main concern is that the Delft3D model has not yet been applied to irrigation canals, and other constraints may be identified after using the Delft3D model. Nevertheless, because of its advantages, the Delft3D model will be tested and its suitability for irrigation systems will be verified after adapting it. In the case where it works properly we will use it to simulate the morphological changes and we can get the benefits of morphology factor predictions which can help the designer and operation planner choose the best canal operation for newly designed canals, and to modify or change canal operation for existing canals. For more details regarding the Delft3D model and its governing equations please refer to these details in chapter1 section 1.7.

The main objective of this chapter is to verify whether the Delft3D model, which represents river networks well, can be used for simulation of sediment transport in irrigation canals.

2.2 METHODS

2.2.1 Model set-up

An important criterion for irrigation canals is the b/h ratio recommended values to be between 3 and 4 (Mendez, 1998), where b is the canal bed width and h the water depth.

For this chapter the following b/h ratios were considered: maximum (4) for wide canals, minimum (3) for narrow canals and in-between (3.5) for medium canals, with rectangular and trapezoidal shapes. The inflow is given to the model in time series. The flow in irrigation canals was assumed to be steady non-uniform flow during the time step. The flow is steady as the flow rates of the outlets do not change with time and is non-uniform as the depth changes with location over the entire canal. For cohesive sediment fractions, the fluxes between the water phase and the bed are calculated with Partheniades–Krone formulations (Partheniades, 1965) for deposition and erosion:

In this chapter C_b^l =2 kg/m^3 for cohesive sediments, M^l =0.0001 kg m^{-2}s^{-1}, τ_{cw} varies along the canals, $\tau_{cr,e}^l$ = 1.8 N/m^2, while $\tau_{cr,d}^l$ = 1000 N/m^2. In case water supply changes the shear stress (τ_{cw}) will change accordingly.

2.2.2 Description of the hypothetical case study

The schematization of the system in this chapter consists of a main canal with a length of 1 km and a branch canal of 0.5 km, which takes water from the middle of the main canal. There are six observation points located at different locations in the main and branch canals, as shown in Figure 2-1.

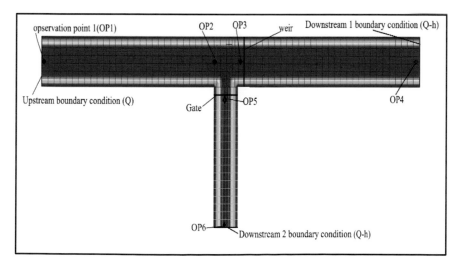

Figure 2-1 Hypothetical case study with all observation points

The following general data of the medium (typical) canals are as follows: The length of the main canal is 1000 m; the length of the branch canal is 500 m; the velocity of main canal is 0.65 m/s; the side slope of trapezoidal canals is 1:1, the designed b/h ratio is between (3 - 4), the depth changes with location over the entire canal, since it is non-uniform flow. Other design criteria are listed in Table 2-1.

Table 2-1 Design criteria for all cases

Scenario	Type of canal	b/h	b-main	H	Q	S₀	Sₛ	N	b-branch	initial WL
1a	Non-wide (rectangular)	3	3	1	1.95	0.0003		0.019	1	32.7
	Medium (rectangular)	3.5	7	2	9.1	0.0002		0.026	2	33.8
	Wide (rectangular)	4	12	3	23.4	0.00012		0.027	6	34.88
1b	Non-wide (trapezoidal)	3	3	1	3	0.0003	1::1	0.021	1	32.7
	Medium (trapezoidal)	3.5	7	2	12	0.0002	1::1	0.028	2	33.8
	Wide (trapezoidal)	4	12	3	29	0.00015	1::1	0.028	6	34.88
2	Medium (rectangular)	3.5	7	2	9.1	0.0002		0.026	2	33.8
	Medium (trapezoidal)	3.5	7	2	12	0.0002	1::1	0.028	2	33.8

b= canal width (m);
h= water depth (m);
Q= discharge ($m^3 s^{-1}$);
S_o= longitudinal slope for canal (-);
S_S= side slope for trapezoidal canals (-);
n= Manning roughness (s $m^{-1/3}$).

The discharge is determined starting from known discharge at the offtake of the main canal (input data) and the discharge which is withdrawn from the main canal by the branch canal (Q_b). Based on the continuity equation, the discharge at the end of the main canal (Q_{out}) should be:

$$Q_{out} = Q_{in} - Q_b \qquad (2\text{-}1)$$

For this reason observation points are located at the beginning of main canal (p1), at the end of main canal (p4) and at the branch canal (p5). According to the continuity equation, the discharge at p5 should equal the difference between p1 and p4.

2.2.3 Scenarios

The first scenario simulates the flow without sediment in rectangular and trapezoidal canals for the different b/h ratios 3, 3.5 and 4, with and without structures. This scenario verifies whether Delft3D can satisfactorily simulate flows in irrigation canals from a hydrodynamic point of view. In the second scenario cohesive sediment will be added to simulations for medium canals with a b/h ratio = 3.5. In this hypothetical case study, a concentration of 20 000 ppm is assumed to be entering the main canal.

2.2.4 Model calibration

To calibrate the model from a hydrodynamic point of view and obtain steady state condition for flow, the results of the Delft3D simulation without sediment are compared to results from DUFLOW modelling. DUFLOW is a program which is used for the simulation of 1D unsteady flow in open canals, for the same canal specifications (following the method described in (Osman et al., 2016).

Double checking for result will be done by using the root square method (R^2) and Nash-Sutcliffe model efficiency (NSE) method. If the results of the two models are close to each other, and if R^2 and NSE around 1, the Delft3D model will be considered adequate for hydraulic simulation in irrigation systems.

2.2.5 Initial conditions

Water level = (32.7 m+MSL (mean sea level) for narrow canals, 33.8 m+MSL for medium canals and 34.88 m+MSL for wide canals).

2.2.6 Boundary conditions

The Delft3D model will be run in steady state conditions according to the field conditions (data assumed). There are two boundary conditions in the main canale: (the upstream boundary is discharge as time series, and downstream boundary is Q-h relationship) while for the branch canal only one downstream boundary condition which is Q-h relationship.

2.3 RESULTS

2.3.1 Scenario 1a: rectangular canals with different sizes and different b/h ratios

Wide rectangular canal with b/h ratio 4.0

The first case considers a wide canal with width of 12 m and water depth of 3 m for the main canal. The branch canal width is 6 m and water depth is 3 m, the designed discharge for this canal is 23.4 m³/s. The graphs in Figure 2-2 show the large similarity between the results obtained from Delft3D and those from Duflow.

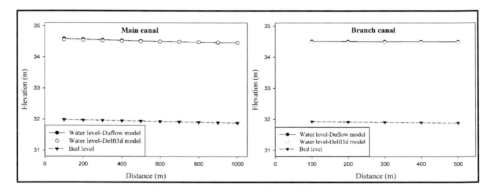

Figure 2-2 Comparison between Delft3D and DUFLOW in both (wide rectangular) canals

The simulated water levels in the upstream part of the main canal differ by 3 cm between the two models. There is no difference in water level in the downstream part. In the branch canal, the difference between model results is 3 cm upstream and 2 cm downstream after calibration. Despite these small differences, the match between DUFLOW and Delft3D is considered as very good.

Double checking the results, the R^2 is 0.99 and the Nash-Sutcliffe efficiencies (NSE) are -1.2 in the main canal, and 0.98 and -7.5 respectively in the branch canal, showing a good match between the results from DUFLOW and Delft3D. Further, at observation points P1, P4 and P5 (branch canal), the discharge equals to 23.4 m³/s, 4.49 m³/s and 18.91 m³/s respectively. This means that continuity equation (discharge at P5 should equal the difference between P1 and P4) is valid.

The velocity distribution resulting from the Delft3D simulation (Figure 2-3) is realistic and conform to hydraulic theory: higher velocities in the centre of the canal and lower velocities at the sides because of wall roughness. In the branch canal the velocity is higher at the outer bend and lower in the inner bend. Near the diversion to the branch canal because of the curvature, we will get extra velocity component due to the centrifugal force.

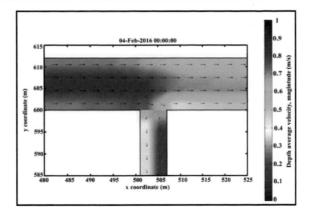

Figure 2-3 Velocity distribution along the system

From these results and observations it was concluded that the Delft3D gives adequate hydrodynamic results for wide irrigation canals with a b/h ratio of 4.

Medium size rectangular canals with medium b/h ratio of 3.5

Simulations in Delft3D and DUFLOW were compared for canals of 7 m width and 2 m water depth in the main canal, and width of 2 m in the branch canal. The difference in water level between the two models was 5 cm upstream of the main canal and 1mm at downstream. In the branch canal the difference was 5 cm upstream of the branch canal and 1 cm downstream. Double checking the results, the R^2 was 0.99 and NSE was 0.95, which means a very good match between the two models. Also a visual inspection of the velocity distribution showed that Delft3D provides realistic simulations for a b/h ratio of 3.5.

Narrow rectangular canals with small b/h ratio of 3

Lastly simulations were compared for narrow canals of 3 m width and 1 m water depth for the main canal, and 1m width and 1m initial water depth in the branch canal. Figure 2-4 compare the results.

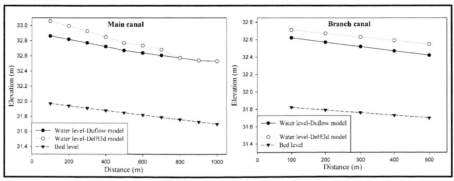

Figure 2-4 Comparison between Delft3D and Duflow in both (wide rectangular) canals

The difference in the water level between the two models is 19 cm upstream of the main canal and 3 cm at the downstream part. In the branch canal, the difference is of 5 cm upstream and 1 cm downstream. The water level simulated by Delft3D model is higher than the water level resulting from DUFLOW. This results seems invalid because in 3D models (such as Delft3D) models friction losses are less than in 1D models (such as DUFLOW) from numerical point of view. The lesser friction losses should lead to lower water level in the 3D model. The big difference in water level can be explained by the sensitivity of the Delft3D model to the side wall friction in narrow canals as the model was originally designed for big scale water systems such as rivers and estuaries. The Delft3D model calculates the roughness for the bed only, since simulations of rivers and coastal shores are not affected by side walls. In this chapter the roughness of walls has been included. This roughness is represented only by Z_o, roughness length, which can be obtained from Nikuradsa (K_s) in the White-Colebrook formula (Deltares, 2016), where:

$$Z_o = K_s / 30 \qquad\qquad (2\text{-}2)$$

K_s can be obtained from the Chezy formula (Deltares, 2016)

$$C = 18 \log (12 \, H / K_s) \qquad\qquad (2\text{-}3)$$

This formula for rivers, estuaries and coastal shores, accounting for water depth and bed but not for the sides walls whose effect on river flow negligible. In irrigation canals where the side walls have considerable impact on the flow, the authors propose using hydraulic radius (R) instead of water depth (H). The principl of Delft3D calculation is done per cell. For each cell with rectangular shape, width=b and water depth=H.

$$R = A/P \quad\longrightarrow\quad R = B*H/B \quad\longrightarrow\quad R = H \qquad\qquad (2\text{-}4)$$

Equation 2-3 becomes:

$$C = 18 \log (12 \, R / K_s) \qquad\qquad (2\text{-}5)$$

To test the sensitivity towards wall roughness (Z_o) in Delft3D, different values of Z_o have been chosen, larger and smaller than the calculated value of 0.0005 m based on the observed R (Table 2-2).

Table 2-2 Difference in water level between the two models given by different values of Z_o

Wall roughness Z_o (m)	Difference- main canal (cm)	Difference-branch canal (cm)
0.01	- 28 *	- 20
0.0001	9	4
0.00005	6	3

* Minus sign means that water level Delft3D model higher than the water level resulted from DUFLOW which is not realistic

The wall roughness $Z_0 = 0.00005$ which is ten times less than the calculated Z_0, is considered the best value giving the smallest difference. That means that the roughness of walls must be much smaller than the canal bed roughness to obtain realistic results in the Delft3D simulations in canals (Figure 2-5). After adjusting the side wall friction also the velocity distribution provided by Delft3D model was realistic. If the velocity is well predicted, this will lead to better prediction of deposition and erosion.

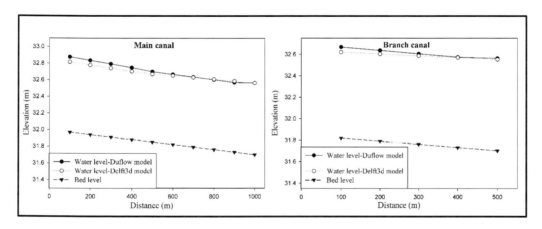

Figure 2-5 Comparison between Delft3D and Duflow models in both (narrow rectangular) canals after adaptation

2.3.2 Scenario 1b: trapezoidal canals with different sizes and different b/h ratios

After the rectangular canal shape gave reliable results, the trapezoidal shape for different canal sizes and b/h ratios were tested. The simulation provided satisfactory results. Figure 2-6 shows the highest velocities in the center of the canal with decreasing velocities towards the side slope because of wall roughness. In the branch canal the velocity is higher at the outer bend and lower in the inner bend. This behaviour of the velocity seems realistic and corresponds to hydraulic theory.

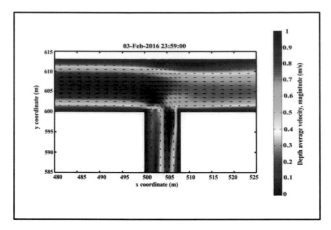

Figure 2-6 Velocity distribution along medium trapezoidal canals

2.3.3 Scenario 1c: canals with structures (rectangular and trapezoidal)

To test whether Delft3D is able to simulate flows in irrigation canals with hydraulic structures, two cases have been chosen: with rectangular and trapezoidal shapes of both medium size (b/h ratio = 3.5) and with an undershot weir in the main canal just after the diversion and a gate at the beginning of the branch   levels in the upstream of the main canal, whereas downstream of the main canal, there is 4 cm difference. The small differences show that the Delft3D simulates water levels well. For the branch canal, the difference between the results upstream is 16 cm and downstream is 12 cm. Even with these small differences, the Delft3D results are still adequate.

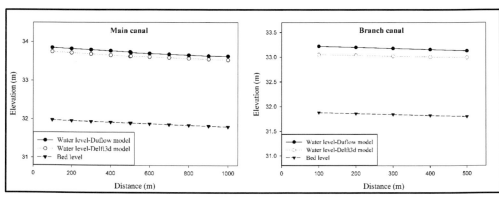

Figure 2-7 Comparison *between Delft3D and Duflow in both (medium trapezoidal) canals*

The simulation results in Figure 2-8 show a realistic velocity distribution with higher velocities in the center of the canal and lower velocities at the sides because of wall roughness. The vertical velocities vary with depth and follow a logarithmic distribution with velocities near the surface higher than near the bottom, except where the water passes over the weir and velocities are equal with depth (Figure 2-12). In the branch canal the velocity is higher at the outer bend and lower in the inner bend, and has its logarithmic distribution downstream the gate. The velocities near the surface are higher than the velocities near the bottom, except at the gate where water pass under the gate and the distribution of water is the opposite where bottom velocities higher than surface velocities and it depends on the gate opening. This behaviour of the velocity is realistic and corresponds to the hydraulic theory.

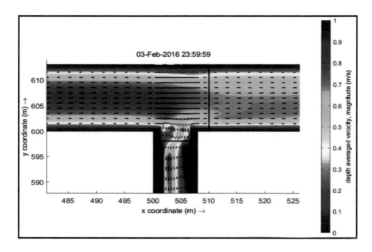

Figure 2-8 Velocity distribution along the system

Rectangular canals

Similarly, the simulation results of flows in canals with rectangular shapes from Delft3D and DUFLOW proved very similar.

Concluding, the simulation results under scenario 1a, 1b and 1c confirm that Delft3D can generate satisfactory results in simulating flows in irrigation systems from hydrodynamic point of view.

2.3.4 Scenario 2: simulations with cohesive sediments

After obtaining satisfactory results from Delft3D for flow simulation in irrigation systems from a hydrodynamic point of view, sediment was added to the first scenario to test Delft3D from a morphodynamic perspective. In practice, sediment concentration in canals varies: some canals have very little or no sediment. Other canals suffer from high concentrations of about 2000 ppm or more since they withdraw their water from rivers which are highly loaded with sediment that reaches $1.4*10^9$ t yr^{-1} (Kondolf et al., 2014). In this scenario, it is assumed that cohesive sediments with a concentration of C = 2000 PPM enter the irrigation system at the main canal. The canal has been chosen as medium size with a b/h ratio of 3.5 (as in scenario 1c).

The model is set to simulate three months using the discharge as an upstream boundary condition and a (Q-h) relationship as the downstream boundary condition at the end of both canals. The initial condition in this scenario is as follows: water level = 34 m, initial sediment concentration = 0, initial layer thickness of sediment = 0.15 m.

In Delft3D, the default value of the settling velocity for cohesive sediment is 0.25 mm/s. However, settling velocity need to be adapted for irrigation systems since it depends on the physicochemical properties of the sediment water system and flow parameters (Partheniades, 1986). There are many formulae that determine the settling velocity of cohesive sediment (Huang et al., 2006) such as (Nicholson & O'Connor, 1986), Burtan *et al.* (1990) and (Van Leussen, 1994). (Krone, 1962) found that the settling velocity increases with the sediment concentration and proposed the following formula:

$$W_s = K\ C^n \tag{2-6}$$

Where:

W_s	= settling velocity (m/s)
C	= suspended sediment concentration (g/l)
k	= empirical constant (-)
n	= an exponential (-)

(Krone, 1962) estimated n = 4/3, k = 0.001, so W_s = 0.001 (20) $^{4/3}$ = 0.05 m/s or 50 mm/s, whereas Cole and Miles (1983) estimated k between 0.001 and 0.002 and n = 1, W_s = 0.0015 (20) = 0.03 m/s or 30 mm/s. In this chapter settling velocity has taken as 30 mm/s as recommended by Cole and Miles (1983) since the maximum value which can be used in the model is 30 mm/s.

To evaluate the modelling results and suitability of Delft3D for simulating cohesive sediment in irrigation canals, important parameters such as velocity, cumulated sediments, bed level, and water level are checked below.

Cumulated cohesive sediment

When the model is run after the introduction of cohesive sediments with a concentration equalling 2000 PPM, the initial layer thickness of sediment in the canal bed starts to be eroded from the beginning of the main canal till 500 m. Just before and after the weir maximum erosion takes place due to the higher velocity at that point in the canal because the weir causes some disturbance to the water flow and the water passes over the weir causing this higher velocity. After some 500 m till the end of the main canal, deposition of sediment occurs because of the reduction in velocity due to the diversion of water to the branch canal. In the branch canal and upstream the gate, because of the high velocity, erosion will occur, whereas downstream of the gate and the rest of the branch canal a deposition will occur. Figure 2-9 presents the deposition and erosion in the cross section of the main canal (with the trapezoidal and rectangular shape) in different positions.

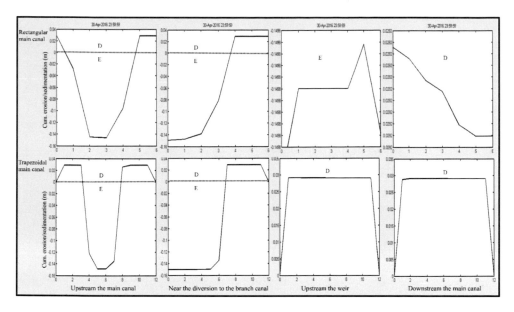

Figure 2-9 The deposition (D) and erosion (E) in different trapezoidal and rectangular cross sections of the main canal

From Figure 2-9 we can see that there is no deposition upstream the main canal, but there is small erosion resulted from eroding the initial sediment layer at the bed. While there is small depositon near the diversion due to the reduction in velocity. Upstream the weir there is significant deposition because of weir effect. A negligible deposition occue at the rest of the main canal. Almost all the sediments moved to the branch canal. The Delft3D results were double checked using the well established Partheniades-Krone formulae (Deltares, 2016) and the simulated results from Delft3D matched the compuation results using this formulae.

Bed level and water level

For both rectangular and trapezoidal canals, because of the erosion which occurred at the beginning of the main canal till 500m, bed level has been lowered and this lowering has affected the water level slightly, while from 500 m till the end of the main canal the deposition of sediment has raised the bed level which has led to a rise in the water level. For more details, see movies S8 and S9 in the supplementary data, where S8 represents the bed level change in a rectangular main canal at and downstream the weir, while S4 represents the bed level change in a trapezoidal main canal at and downstream the weir. The link for supplementary data is https://drive.google.com/open?id=1AZMRlyArXQmR2GUnBLRYBuPEGE5yaIuA.

For the branch canal, the deposition of sediments along the branch canal has raised the bed level which leads to raise the water level. Except for upstream and downstream the gate, the bed level has been lowered due to erosion and has led to a reduction in the water level. Figure 2-10 presents the change in the morphology of the bed within the cross section of the trapezoidal main and branch canals in different positions. The same results for rectangular canals have been obtained.

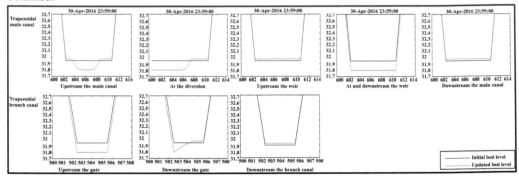

Figure 2-10 The bed morphology development within the cross-section of the (main and branch) trapezoidal canals in different locations

Velocity

The velocity distribution is logarithmic with higher velocities in the top water layer while lower velocities are found near the bed along the main canal except above the weir where water flows over it, and its distribution changed to become uniform where the top and bottom layer velocities are equal (Figures 2-11 and 2-12).

In the branch canal velocity distribution is also logarithmic, with higher velocities in the top layer of water and lower ones near the bed along the branch canal. Except for near the gate, water flows under the gate and the velocity distribution changed where top layer velocity became less than bottom layer velocity (Figures 2-11 and 2-12).

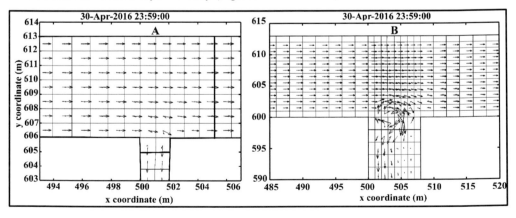

Figure 2-11 Velocity distribution in the system (A- The rectangular canal and B- The trapezoidal canal)

While Figure 2-11 shows that the higher velocities are found in the center of the canal, and at the sides there are lower velocities because of wall roughness, in the trapezoidal system there is circulation because of the side slope. The velocity shown in Figure 2-11 is at the end of the simulation time, while the Supplementary Data of this chapter demonstrate the velocity distribution along the simulation period of the analysis.

For more details, see movies S1 and S2 in the supplementary data, where S1 represents the Velocity in the medium trapezoidal canal-without structures, while S2 represents the Velocity in the medium trapezoidal canal-with structures. The link for supplementary data is: https://drive.google.com/open?id=1AZMRlyArXQmR2GUnBLRYBuPEGE5yaIuA.

In the branch canal the velocity is higher at the outer bend and lower at the inner, and this behaviour of the velocity is realistic and corresponds to the hydraulic theory.

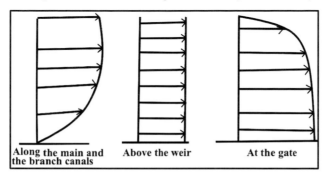

Figure 2-12 Velocity distribution in the system (A- The rectangular canal and B- The trapezoidal canal)

Figures 2-13 and 2-14 show the different patterns of velocity distribution in the rectangular and trapezoidal main canals at different positions where, as mentioned above, the velocity in the top layer is higher than that at the bottom; higher velocities also exist downstream of the weir.

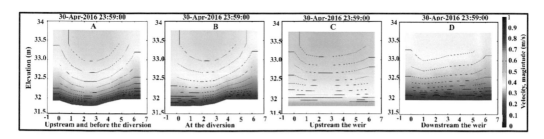

Figure 2-13 Depth averaged velocity within the cross section of the rectangular main canal (A- upstream and before diverting to the branch, B- at the diversion, C- upstream the weir, D- downstream the weir)

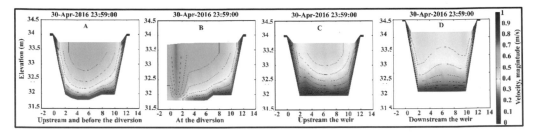

Figure 2-14 Depth averaged velocity in a cross section in the trapezoidal main canal (A-upstream and before diverting to the branch, B- at the diversion, C- upstream the weir, D-downstream the weir)

The velocity distribution shown in Figures 2-13 and 2-14 is at the end of the simulation time, while the Supplementary Data of this chapter demonstrate the velocity distribution in the main canal along the simulation period of the analysis. For more details, see movies S3 and S4 in the supplementary data, where S3 represents the velocity in the medium rectangular canal, while S4 represents the Velocity in the medium trapezoidal canal. The link for supplementary data is

https://drive.google.com/open?id=1AZMRlyArXQmR2GUnBLRYBuPEGE5yaIuA.

Figure 2-15 shows the different patterns of velocity distribution in the trapezoidal branch canal at different positions. While the higher velocities exist upstream of the gate and at the gate and downstream of the gate the velocity starts to adjust itself, after the gate for the rest of the branch canal the velocity distribution became logarithmic, with higher velocities in the top layer of water and lower velocities near the bed.

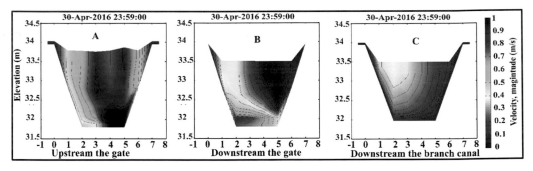

Figure 2-15 Depth averaged velocity in a cross section in the branch canal (A- upstream the gate, B- downstream the gate, C- Downstream the branch canal)

The velocity distribution shown in Figure 2-15 is at the end of the simulation time, while the Supplementary Data of this chapter demonstrate the velocity distribution in the branch canal along the simulation period of the analysis.

For more details, see movies S5, S6 and S7 in the supplementary data, where S5 represents velocity in branch canal upstream the gate, while S6 represents velocity in branch canal at the gate and S7 represents the velocity in branch canal downstream the gate. The link for supplementary data is:

https://drive.google.com/open?id=1AZMRlyArXQmR2GUnBLRYBuPEGE5yaIuA

2.4 DISCUSSION

2.4.1 Adapting Delft3D for irrigation systems

After calibrating the Delft3D model for a hypothetical irrigation canal system and running it for 3 days per simulation under different scenarios, several issues came up that needed adaptation in the model formulation: first, to accommodate the (Q–h) relationship at the downstream boundary condition; second, to deal with side wall friction; and third, to deal with the long running time. These are discussed in detail below.

Q–h relationship

In hydrodynamic models used for canals (such as Duflow), the Q–h relationship is usually used as the downstream condition. In irrigation canals it is not correct to fix the water level as the downstream boundary condition because in this way the model will be forced to have this defined water level at the end of the canal and this is difficult in view of the different water requirements for branch canals. Hence the Q–h relationship is the preferred downstream boundary condition. However, in the Delft3D model, this option proved a challenge because in the Delft3D model this relationship is reversed. It is formulated as an h–Q relationship at the downstream boundary. At the same time in Delft3D it is not possible to present a zero discharge or zero water level. For rivers and estuaries this does not pose a problem (it is unlikely that they will run dry if there is water upstream). In irrigation canals it is not uncommon that all water from upstream is diverted to branch canals or fields and canals run dry downstream. In the Delft 3D computations, a zero discharge is considered as a dry cell and the model run is aborted.

To solve this difficulty and overcome the dry bed condition in Delft3D, the bed levels of the last two grid cells at the downstream end of the canal were lowered. This means creating a virtual drop structure at the downstream end of the canals, to accommodate the sudden change in elevation and to dissipate the energy without causing scouring in the canal itself. This virtual drop structure should be in the range 0.3–0.8 m to lower the water level while allowing a subcritical flow in the main canal and avoiding the critical flow. In this case chapter a drop of 0.5 m was found to be adequate.

This hydraulic trick did not affect the overall result of the model simulation. To double-check also in the Duflow simulations the level in the last cross section was lowered by 0.5 m without affecting the overall modelling results.

Wall roughness

The simulation results show that the Delft3D model is very sensitive to wall roughness, especially for small rectangular canals with a b/h ratio smaller than 3. The results for trapezoidal canals show better performance than the rectangular canals because the top section in the trapezoidal canals is much wider than in rectangular canals and hence the impact of side walls is less. This chapter found that for small rectangular and trapezoidal canals satisfactory results were obtained if the wall roughness was set an order of magnitude lower than the bed roughness. Further work is needed to investigate how the adapted roughness settings in the model affect the simulation of the cohesive sediment, in particular in small rectangular canals.

Simulation time

The computation time of a simulation run in Delft3D is several orders of magnitude longer than in Duflow. Table 2-3 shows the running times for a simple system composed of the main canal with one branch canal and several structures, under different scenarios, using the newest version of Delft3D 4:02:03.

To reduce running times, a tool in Delft3D called 'domain decomposition' was used. Domain decomposition splits the domain into two or more domains and then compiles these domains. Although this reduced the computation time by 60% it remains long. The running time is further reduced by using a refinement property in Delft3D called 'flexible mesh'. This allows the user to choose different grid sizes for different locations along the system. In long straight stretches of the canal with few hydrodynamic changes a large grid cell is sufficient, while at bends, structures and offtakes a smaller grid cell is needed to properly reflect the variations in hydrodynamics and sediment deposition. This property reduces the number of computational grid cells and thus leads to a reduction in computation time. Obviously, computer efficiency and specifications play a big role in computing time. In this chapter two laptops were used. One has very simple specifications (dual core Hp proBook 6570b) and the other is a higher-performance computer is (quadrilateral core Hp Z Book15 G3); the latter reduced the simulation time by 40%.

While computation time has been a challenge in this chapter, it is envisaged that this issue will reduce because laptops continue to become faster, and several improvements in the Delft3D model to reduce computational requirements are being implemented. Lastly, supercomputers are becoming more widely accessible and affordable.

Table 2-3 Simulation time for different cases

Case (without sediment)	Time (h)[a]	Time (minutes) with domain decomposition and after refinement
Non-wide rectangular canals	36	120
Medium rectangular canals	21	90
Wide - rectangular canals	14	45
Non-wide- trapezoidal canals	45	135
Medium- trapezoidal canals	26	100
Wide- trapezoidal canals	20	55
Case (with sediments)	Time (Days)	Time (Days) with domain decomposition and after refinement
Medium-rectangular canals – Cohesive	5.5 Days	2 Days
Medium-Trapezoidal canals – Cohesive	7 Days	3 Days

[a] If better computer are used, these computation times would be 40% shorter

2.5 CONCLUSIONS

This chapter applied Delft3D for the simulation of different scenarios in a hypothetical irrigation canal set-up consisting of a main canal, branch canal and two structures (a weir in the main canal and a gate in the branch canal). While the Delft3D model is mostly used in rivers and estuaries, in this chapter the authors chose to apply it to an irrigation setting for two main reasons. First, Delft3D is one of the few models capable of simulating cohesive sediment (i.e. sediment in suspension). So far, most of the sediment transport studies in irrigation canals have been conducted for non-cohesive sediment (i.e. bed load). Second, Delft3D is capable of 2D or 3D simulation. The few existing studies of cohesive sediment in irrigation canals use 1D models (such as SIC and FSEDT). This may be adequate for non-cohesive sediment but will be too coarse for application of cohesive sediment that floats in different directions along the canal and in cross sections, in particular at bends, offtakes, and structures.

After adaptations in the model, the initial results of using Delft3D in irrigation canals were very promising. Comparing results with Duflow simulations, it is concluded that Delft3D provides good results in simulating water levels in the main and branch canal from a hydrodynamic point of view. Further, the Delft3D model was able to provide a realistic image of velocity distribution along the system and in canal cross sections. Lastly, the Delft3D model provided realistic results for cohesive sediment behaviour and transport, in both horizontal and vertical directions including deposition and uptake (erosion) at the water–bottom interface. It realistically showed how cohesive sediment moves and is distributed along the canals. Therefore, we conclude that also from a morphodynamic point of view Delft3D is capable of simulations in irrigation systems.

Some challenges remain, however, such as the sensitivity of side wall roughness in small rectangular canals where in order to obtain a reasonable output for the Delft3D model as compared with Duflow model outputs, we need to adjust and reduce the wall roughness which will add uncertainty to sediment transport prediction (just for small rectangular canals). The other challenge is the expected long computation times in more complex canal networks (with main canals, multiple secondary and tertiary canals, with bends and multiple structures) than those explored in this chapter. From the initial results we conclude that these challenges can be adequately addressed in further studies and with ongoing adaptation of the Delft3D model. This chapter provides a basis for further work and shows the importance of doing so for the simulation of cohesive sediment in irrigation canals.

3

THE USE OF 2D/3D MODELS TO SHOW THE DIFFERENCES BETWEEN COHESIVE AND NON-COHESIVE SEDIMENTS IN IRRIGATION CANALS

This chapter has been submitted to the American Journal of Irrigation and Drainage Engineering (ASCE) and currently under review:

Theol, Shaimaa, Bert Jagers, F.X. Suryadi, Charlotte de Fraiture. The use of 2D/3D models to show the differences between cohesive and non-cohesive sediments in irrigation systems.

The use of 2D/3D models to show the differences between cohesive and non-cohesive sediments in irrigation canals

ABSTRACT

Sediment deposition in irrigation systems results in disruption of water distribution and high maintenance costs. The vast majority of studies on the behaviour of cohesive and non-cohesive sediments have been done in rivers and estuaries. The relatively few studies on sediments in irrigation systems deal with non-cohesive sediments mostly using 1D models. In practice, irrigation systems that tapping from natural rivers often face a mix of cohesive and non-cohesive sediments. The sedimentation patterns usually are non-uniform, especially around offtakes and structures. Therefore, the authors used Delft3D, a hydro-dynamic 2D/3D model, and adapted it for use in irrigation canals to test different scenarios of pure and mixed sediments with varying concentrations and discharges. The authors found that Delft3D was able to provide additional insights in the behaviour of cohesive and non-cohesive sediments in irrigation canals and showed the importance of using 2D/3D models. Also, it is found that cohesive sediments are more sensitive to the variations in discharge and velocity compared to non-cohesive sediments. Simulations reveal that where non-cohesive sediments are present in a mixture with cohesive sediments, the deposition is slower than in the case of pure non-cohesive sediments of the same concentration.

3.1 INTRODUCTION

The deposition of sediments in irrigation systems causes a range of problems, such as the reduction of conveyance capacity, the blocking of offtakes and gates, canal breaching, bank erosion and errors in the calibration of measurement structures. Accumulation of sediments raises the canal bed and the water level, leading to disturbances in water distribution. This may result in some fields receiving more and other fields less water than planned, leading to undersupply, inequity, and to decline of the irrigated area, and ultimately reduced crop production (Munir, 2011; Paudel, 2010). Sediments not only affect irrigation performance but also have a considerable impact on the requirements in terms of efforts and money to keep irrigation systems in running condition, consequently potentially affecting their (financial) sustainability. Consequently, sediment management in irrigation systems receives substantial attention from irrigation managers and scholars.

Cohesive or fine sediments are generally transported in suspension and are primarily composed of clay-sized material with strong inter-particle forces due to their surface ionic charges. The inter-particle forces are the dominant factor in their behaviour. Non-cohesive or coarse sediments, transported as bed material, are mainly composed of sand and fine gravel material with weak inter-particle forces, where in their motion, non-cohesive particles behave independently from each other, except in very high concentrations.

Cohesive sediments can pose serious problems in irrigation systems, especially in high concentrations, where accumulation occurs in so-called flocs. These flocs are difficult to remove and can cause obstruction of the water flow, rising of water levels and increased growth of weeds. In irrigation canals with limited flows and low water depths these consolidated flocs are difficult to remove or brought back into suspension (Lawrence, 1998). These aspects lead to a reduction in conveyance capacity and an increase in maintenance required. Sediment control structures, constructed at the head of irrigation canals, typically trap coarse sediments like sand and gravel but are not effective in trapping cohesive sediments.

The behaviour of non-cohesive sediments in irrigation canals under different operation scenarios are relatively well studied, for example, (Jinchi et al., 1993; Paudel, 2010; Sherpa, 2005) for systems in Nepal; Sutama (2010) for Indonesia and Munir (2011) for Pakistan. The behaviour of cohesive sediments in irrigation canals is far less studied. Most of these studies focus on the quantification of cohesive sediments in canals and their impact on the performance (Belaud & Baume, 2002). An example of one of the few studies on the effect of canal operation on cohesive sediment behaviour is the work by Osman et al. (2016) on the Gezira Scheme in Sudan.

The impact of silt deposition on the aggradation of the irrigation canals in Zitny Ostrov was evaluated for a period of 11 years by Dulovičová and Velísková (2009). It is stated that the thickness and structure of the silt are factors influencing the groundwater interaction. The silt has increased about 45.5%, Dulovičová and Velísková (2009) found that the knowledge of the actual physical silt thickness in the canal is an important element for the future study of the interaction between the canal networks and groundwater.

The mathematical models that were used to simulate sediment behaviour in irrigation systems are mostly one dimensional (1D). 1D-models simulate the longitudinal direction of the sediment erosion and deposition, assuming that the sediment is moving only in the longitudinal direction. 1D models are unable to predict the sediments in the lateral direction. In reality, sediment deposits in irrigation canals are unevenly distributed, in particular near off-takes and structures, especially in the case of cohesive sediments. Further, the behaviour of non-cohesive and - to a lesser extent - cohesive sediments in irrigation canals have been simulated in isolation. Many irrigation canals withdraw water containing mixed sediment (cohesive and non-cohesive). In Sudan, in the Gezira scheme, there is mixed sediment with 70% cohesive sediments and 30% non-cohesive sediments as stated by Osman (2015), however, Osman considered the cohesive sediments only. Other researchers consider non-cohesive sediments only, due to the complex behaviour of sediment mixture. In this research, we aim to address this by introducing a new method considering the mixed sediment (both cohesive and non-cohesive).

Therefore, the scope of this chapter is to use a 2D/3D model to demonstrate the differences of behaviour and the differences of erosion/sedimentation patterns between cohesive and non-cohesive sediments in irrigation canals, studying the sediment type impact on irrigation canal management you need to know the differences in behaviour. To contrast the behaviour of cohesive and non-cohesive sediments in irrigation canals, as well as their interaction, the authors selected the Delft3D model. This 2D/3D model includes the process formulations for both types of sediments and has a proven track-record of simulating sediment behaviour. The Delft3D model has mostly been used for rivers and estuaries (Gebrehiwot et al., 2015; Van der Wegen et al., 2011). To the authors' knowledge, 2D/3D has been used very limited for irrigation canals. The difference in behaviour between cohesive and non-cohesive sediment is well-known from fundamental research and river applications, but this chapter puts these differences in the context of irrigation canals.

3.2 METHODS

3.2.1 *Modelling using Delft3D Governing equations in the Delft3D model*

When designing canal distribution systems, two elements are important. The first element concerns operational or hydraulic aspects, as water flows become non-uniform due to changes in water requirements and variable gate operations to fulfil water demands and keep water levels as required to reach farmers' fields. The second element concerns the sediment transport, as the changes in water flow in time and space are faster than changes in the morphology of canals (Depeweg & Méndez, 2007). The Delft3D-flow solves the Reynolds averaged Navier Stokes equations, provides the hydrodynamic basis for morphological computations and calculates non-steady and steady flow, in addition to transport phenomena (Deltares, 2016). The sediment transport and morphology module supports both bedload and suspended load transport of non-cohesive sediments and suspended load of cohesive sediments.

For cohesive sediment fractions, the fluxes between the water phase and the bed are calculated with Partheniades-Krone formulations Partheniades (1965) for deposition and erosion (Deltares, 2016).

In this chapter the maximum concentration was assumed to be (C_b^l = 3 kg/m^3 or 3000 ppm) for cohesive sediments; which are relevant for irrigation systems like the Gezira scheme in Sudan (Osman, 2015). For example, the maximum concentrations of the cohesive sediment in the Gezira scheme could be as high as 7000 ppm and normally fluctuates between 3000-4000 ppm (Osman, 2015). Based on the Krone (1962) formula the settling velocity (Ws) corresponds to (3000 ppm) equals to 0.12 mm/s. The value of the critical shear stress for erosion ($\tau_{cr,e}^l$) is estimated at 1.8 N/m^2. This value lies in the middle of the range of calculated shear stress 2.3 N/ m^2 (only deposition) and 1.4 N/ m^2 (only erosion), to ensure that both erosion and deposition can take place. For the erosion parameter M ^{1} the default value of 0.0001 kg m^{-2}s^{-1} is used. Although the model equations allow for the specification of critical shear stress for deposition, various researchers (Chan et al., 2006; Sanford & Halka, 1993; Winterwerp & Van Kesteren, 2004) indicate that it does not exist in nature. The authors therefore used default value t$_{cr, d}$ = 1000 N/m^2. The high value for t$_{cr, d}$ causes $S(\tau_{cw}, \tau_{cr,e}^l)$ to be effectively equal to 1. The computation of the suspended sediment transport is done based on solving the advection-diffusion equation numerically (Huang et al., 2008).

For the computation of the behaviour of non-cohesive sediments, the Delft3D model applies the approach developed by Van Rijn (1993). Van Rijn (1993) predicts sediment transport as bed-load and suspended load. A reference height (a) is used to differentiate between these loads; the sediments which move below this reference height are treated as bedload transport and above it as suspended-load transport. The layer situated directly above the Van Rijn reference height is called the kmx-layer. The sediments in this layer which move between the bed and water flow are modelled using sink and source terms. The quantity of sediments entering the flow due to upward diffusion from the reference height, as well as the quantity of sediments dropping out of the flow due to sediment settling, are modelled each half time-step through the source and sink terms.

The advection-diffusion equation solves the sink term implicitly, whereas the source term is solved explicitly. The concentration and concentration gradient at the bottom of the kmx-layer needs to be approximated, in order to determine the sink and source terms. The authors assume a standard Rouse profile between the reference height (a) and the centre of the kmx-layer. For more details regarding the Delft3D model and its governing equations please refer to these details in chapter1 section 1.7.

3.3 MODEL SETUP

3.3.1 *Grid construction, bathymetry, and other parameter assumptions*

The general characteristics (main and branch canal dimensions, sediment characteristics and discharge) are based on an existing irrigation system in Sudan. For the purpose of this chapter, the authors use a simplified system to focus on the basic processes. To explore the behaviour of sediments in a simple but representative canal set-up, the authors constructed a grid for the main canal of one kilometer and a branch canal of 500 meters, with six observation points as depicted in Figure 3-1, (for more details some results for different scenarios in the observation points are presented in the supplementary document). The grid contains 508 cells and 12 cells in the M and N-direction respectively for the main canal. In the branch canal, there are 250 cells in the N-direction and 8 cells in the M-direction. For the 3D simulations, we have used five equidistant sigma-layers.

The authors followed the grid quality criteria of Delft3D with the orthogonality = 0 (i.e. cells are perpendicular to each other, reducing the Courant number that causes the simulation instability), and the smoothness = 1.2 for both M and N directions. To reduce the computation time, the grid is split into the main grid domain and branch grid domain. Simulation results of both domains are compiled using the domain decomposition tool (DD), reducing simulation time to 40% of the taken time to simulate the whole domain (without splitting).

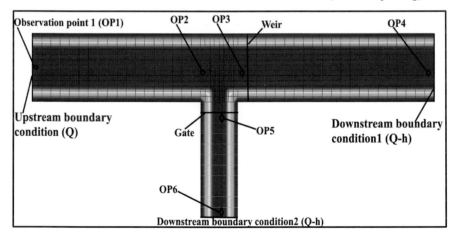

Figure 3-1 Set-up of the hypothetical case with the six observation points

The assumed geometric data are presented in Table 3-1. The irrigation canals which are relatively small and the effect of side-wall is significant. For this reason, in the design of irrigation canals, the b/h ratio between 3 and 4 is recommended. (Mendez, 1998). In this chapter, the authors assumed a canal with b/h of 3.5, with a trapezoidal shape (for more details, the authors have tested different canal sizes with different b/h ratios of 3 and 4, see supplementary document). Based on the side slope, the b/h ratio and the geometric dimensions, the designed discharge for the main canal was calculated to be 7.85m^3/s which is the upstream boundary condition. Furthermore, the flow was assumed to be steady during the time step, as the flow rates of the outlets do not change with time and it was assumed to be non-uniform because the water depths change with the location in the canal.

Table 3-1 Geometric data

	Main canal	Branch canal	Units
Canal length	1000	500	m
Canal width	7	2	m
Canal height	3	3	m
manning	0.028	0.028	s.m$^{-1/3}$
Slope	0.0002	0.0001	---
Side slope	01:01	01:01	---
Structures		Gate fully opened	

Sediments concentration in canals varies with time and location. Some canals have very little or no sediments while others suffer from high concentrations throughout the year or in certain months. In this chapter, the author tested different scenarios with concentrations varying from 100-3000 ppm for cohesive and 30-1000 ppm for non-cohesive sediments. The sediment transport is calculated using the Krone (1962) formula. Parameter values regarding non-cohesive sediments are: D_{50}= 250 μm (coarse sand) with specific density of 2650 kg/m^3. The authors tested several sediment sizes (D_{50}), the results revealed similar behaviour for other sizes (for more details, the authors have tested different particle sizes, see supplementary document). For the transport of non-cohesive sediments, the (Van Rijn, 1993) formulae was used.

3.3.2 Model runs

The authors ran the model for three months using a time-step of 0.9 seconds and a morphological factor (MF) of 10, using both 2D and 3D modes. The results of the 2D and 3D simulations look very similar and identical almost everywhere, but differences occur near the offtake where the flow field is in fact 3D. Where running the model in 2D mode gives a better representation of the sediment transport and of the erosion/sedimentation patterns with acceptable running time while running the model in 3D mode cost more time but gives more information and details about the sediment transport and processes near structures in the vertical direction (Theol, S. et al., 2019a), in this chapter, the graphs are based on the 3D simulations. The small time-step is chosen to avoid the Courant number exceeding 1.0. In mathematical models, the high value of the courant number can destabilize the model. The morphological acceleration factor (MorFac or MF) is an approach to speed up the calculations of the changes in the bed morphology without affecting the simulations. The MF enables the computation of sediment transport and morphological change simultaneously with the hydrodynamics. In this study, the authors used MF=10 to accelerate the computation of bed morphology changes by 10 times for each time step. Consequently, simulating the effective morphological changes over 3 months requires only a simulation period of 9 days, which takes 3 days actual simulation time (CPU time) for 3D model simulations and 1.5 days for 2D simulations. The MF approach simplifies the model setup and operation in comparison with other approaches (Li, 2010).

The initial conditions are set as follows: discharge = 0 m^3/s and water level = 34 m + MSL (the bed level =32 m+ MSL at the beginning of the canal (upstream), calculated water depth is 2 m. Based on this we gave water level as 34 m+MSL). Both the main canal and the branch canal are affected by cohesive and non-cohesive sediments. Initial sediment concentrations for both cohesive and non-cohesive sediments are 0 kg/m^3 while the initial sediment layer in the canal is 10 cm. The boundary conditions are as follows: upstream of the main canal the discharge equals 7.85 m^3/s. The two downstream conditions are the Q-h relation at the end of the main canal and the Q-h relation at the end of the branch canal, both Q-h relations are presented in the supplementary data.

Building on earlier study (Theol, S. et al., 2019b), the authors ran the Delft3D model first without sediments in order to get steady-state flow conditions from a hydrodynamic point of view and to validate some important flow parameters such as velocity, water levels and bed shear stress (1.8 N/m^2), which play a key role in deposition and erosion of cohesive sediments. Hydrodynamic results were validated using the DUFLOW model (Theol, S. et al., 2019b). In the current chapter, the authors compare the differences in behaviour between cohesive and non-cohesive sediments and their interaction by running scenarios with different sediment concentration and flow discharges.

3.3.3 Scenarios

There are many factors that can affect the sediment behaviour and sediment deposition, any change in the system will have an impact e.g. the variation in sediment transport parameters, sediment properties, variations in channel geometry parameter settings and so on. This chapter focus on the variation in the external boundary forcing (upstream discharge and upstream concentration). Additional scenarios covering the variation in grain size (D_{50}) and channel width are included in the supplementary material. To understand the behaviour of and the interaction between cohesive and non-cohesive sediments in different concentrations, the authors formulated a range of scenarios regarding concentration and mix of sediment types, as shown in Table 3-2:

To show the sensitivity of the change in sediment concentrations, scenario 1 simulates variable sediment input, representing low and high concentrations. On the other hand, to show the sensitivity of the change in flow characteristics, scenario 2 simulates variable flow input, representing low and high flows in the canals, mimicking varying crop water demand during the growing season. To understand the differences in sediment behaviour, scenario 3 presents the comparison between pure cohesive sediments with pure non-cohesive sediments. Scenario 4 provides insights into the interactions between the sediment types in different concentration ratios. Further parameter changes to the grain size can be found in the supplementary material.

Table 3-2 Scenarios with different sediment concentration and mixes

Scenario	Simulation	Cohesive bnd. (ppm)	Non-cohesive bnd. (ppm)	%Q "relative to Q_d=7.85 m^3/s"
1- Using different concentrations	1a	N/A *	30	100%
	1b	N/A	100	100%
	1c	N/A	300	100%
	1d	100	N/A	100%
	1e	300	N/A	100%
	1f	1000	N/A	100%
2- Applying different flow discharges as % of (Q_d)	2a	1000	N/A	50%
	2b	1000	N/A	100%
	2c	1000	N/A	150%
	2d	N/A	300	50%
	2e	N/A	300	100%
	2f	N/A	300	150%
3-Pure sediments with extreme concentration	3a	3000	N/A	100%
	3b	N/A	1000	100%
4-Non-cohesive versus mixed sediments	4	100	30	100%

* N/A means that sediment fraction is not included in the simulation

3.4 RESULTS

3.4.1 Impact of variable sediment concentrations

Non-cohesive sediments

Generally, increasing the concentrations leads to sediment build-up and raised canal bed levels. However, the pattern of deposition differs depending on sediment type. Increasing the non-cohesive sediment concentration in the range of 30 to-300 ppm leads to increased deposition in the main canal and hence increased bed level, reduced water depth, reduced cross-sectional area, and increased velocity. Similarly, in the branch canal increased non-cohesive sediment concentration leads to deposition though the effect is less pronounced (Figure 3-2). Low concentrations give minor changes to the cross-section profile while in case of high concentrations (C > 300 ppm), the change in the cross-section profile is more substantial (up to 47% reduction in the cross-sectional area in the upstream of the main canal, while downstream of the main canal the reduction is only 1%, for more details, see figure S 16 in the supplementary document).

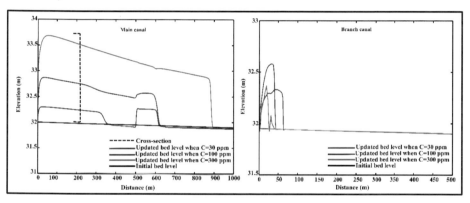

Figure 3-2 Bed level in main and branch canals with different concentrations of non-cohesive sediments, the cross section will be presented in Figure 3-6 right panel.

In the case of increasing non-cohesive sediment concentrations, the flow velocity in the middle of the main canal increases rapidly. The side close to the branch canal has lower velocity due to the higher friction caused by the increasing sediment concentration. The lower velocity leads to the accumulation of non-cohesive sediments, preventing water flow to the branch canal (Figure 3-3).

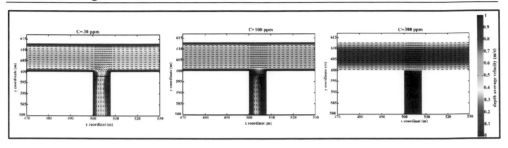

Figure 3-3 Velocity with different concentrations of non-cohesive sediments (C=30, 100 and 300) ppm.

Cohesive sediments

Increasing the cohesive sediments concentration has limited impact on the main canal bed morphology, while the impact in the branch canal is more visible (Figure 3-4), if we would use a higher concentrations for cohesive sediments, the amount of sedimentation is bigger than in case of low concentrations but the patterns are similar, which means cohesive sediments are not very sensitive to the change of concentrations compared to non-cohesive sediments, (for more details, see figure S 17 in the supplementary document). In the case of small concentrations, the erosion occurs upstream of the diversion while for the high concentrations no erosion occurs.

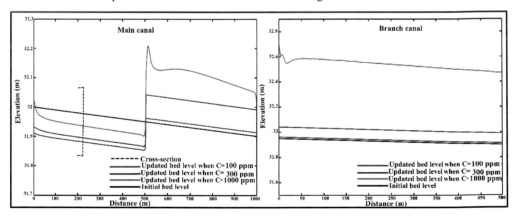

Figure 3-4 Bed level in main and branch canals with different concentrations of cohesive sediments, the cross section will be presented in Figure 3-6 left panel.

In the case of cohesive sediments, the reduction in the cross-sectional area in the upstream part of the main canal is less than the reduction in the case of non-cohesive sediments, while the opposite occurs at the downstream part of the main canal. At the upstream of the main canal, the cross-section profile is reduced by 12%; in the downstream, the reduction is 6%.

In the case of increasing cohesive sediment concentrations, the flow velocity in the main canal increases gradually. The branch canal sees sediment accumulation but is not blocked (Figure 3-5).

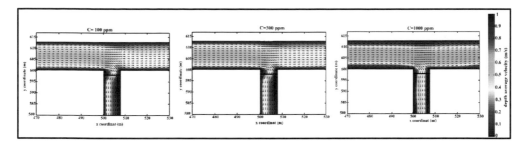

Figure 3-5 Velocity with different concentrations of cohesive sediments (C=100, 300 and 1000) ppm.

The dark brown colour in Figures 3-3 and 3-5 indicates areas with near-zero water flows and very low velocities because of the large sediments accumulation, in particular in the main canal close to the diversion to branch and at the outer bend of the branch (for more details, please refer to the supplementary data see video S1A, video S1B, and video S1C regarding velocity with different non-cohesive concentrations, and see video S2A, video S2B, and video S2C regarding velocity with different cohesive concentrations).

Varying the sediments concentrations has a large impact on the deposition and erosion of sediments, in particular for non-cohesive sediments. The impact is less pronounced for cohesive sediments. The behaviour of non-cohesive sediments is highly sensitive to the change in sediments inputs in comparison to cohesive sediments. To visualize this effect a 2D or 3D representation is required.

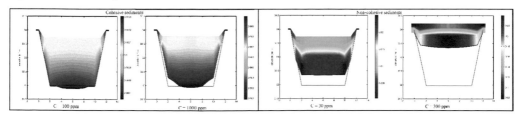

Figure 3-6 Bed level and sediment concentration within the cross-section, right panel with different concentrations of non-cohesive sediments and left panel with different concentrations of cohesive sediments.

The use of 2D/3D models to show the differences between cohesive and non-cohesive sediments in irrigation canals

The different impact of different sediment concentrations on cross-sections are shown in Figure 3-6. The amount and the location of deposition/erosion within the cross-section can be clearly seen. This reveals the importance of 3D representations since a 1D representation would not be able to show that.

3.4.2 Variable flow discharges (Q)

Flows in irrigation canals can vary considerably throughout the season depending on irrigation requirements (which depend on the crop, rainfall, temperature, number of farmers irrigating simultaneously, etc.). Variable flows have a considerable impact on sediment behaviour, depending on the sediment type, Scenario 2 mimics this situation. In the case of cohesive sediments, there is a relationship between the discharge with the shear stress value (Krone, 1962).

Cohesive sediments

The shear stress value is the key factor determining deposition or erosion (Krone, 1962). In the Delft3D model, using the Krone formula, erosion occurs when the bed shear stress exceeds the critical value of 1.8 N/m^2. Deposition occurs when the shear stress is lower than the critical value (Figure 3-7).

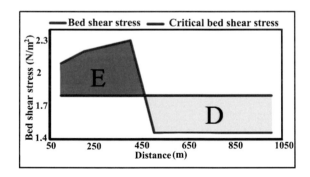

Figure 3-7 Effect of critical shear stress on deposition and erosion in case of cohesive sediments (pink area represent erosion and yellow area is deposition)

Increasing the water supply in the canal increases velocity, leading to increase shear stress, which - if exceeding the critical value - leads to erosion of the already deposited cohesive sediments. This happens at the beginning of the main canal when the water supply is increased by 50% compared to the reference discharge (i.e. 150% Q_d). After the diversion to the branch canal at 500 meters, the bed shear stress decreases to below the critical shear stress and some deposition occurs, although overall the bed level still remains under the initial bed (Figure 3-8 left panel).

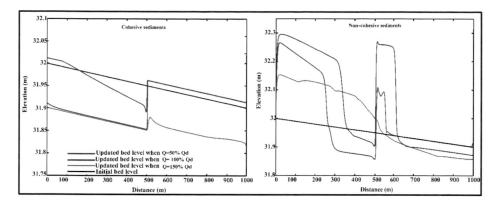

Figure 3-8 Effect of different water supply for cohesive and non-cohesive sediments on the bed level.

Decreasing the water supply to 50% Q_d leads to reduce flow and shear stress, which leads to deposition. However, since the concentrations are low, the resulting bed level is still below the initial bed level for the part of the main canal upstream of the diversion (Figure 3-8 left panel). Downstream of the diversion, the deposition will occur. For high concentrations, the reduced flow has limited impact in comparison to the full supply scenario (Q_d) and the deposition in the main canal is nearly the same (for more details, see figure S18 left panel in the supplementary document). In the branch canal, since the bed shear stress is less than the critical shear stress, there is always deposition and no erosion occurs. Reducing the discharge to 50% leads to more deposition than in the case of Q_d, while increasing discharge to 150% will reduce the deposition rate and increase carrying capacity, therefore flushing the sediments away to the field.

In the case flow discharge is reduced by half, the velocity drops to 0.45 m/s from 0.55 m/s in case of full supply (Q_d). The small deposition hardly affects the cross-sectional area and the water enters the branch canal from the inner bend. When the discharge is increased by 50%, the flow becomes turbulent and the water enters from both bends (Figure 3-9). Because there is hardly deposition or erosion in the main canal, there is no change in the cross-sectional area.

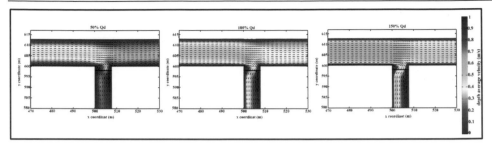

Figure 3-9 Effect of different water supplies discharges (50%Qd, Qd, and 150%Q d) for cohesive sediments on velocity.

Non-cohesive sediments

Reducing the Q_d by half results in the deposition of almost all non-cohesive sediments in the upstream of the main canal. The deposition in the case of 50% Q_d is higher than in the case of Q_d and 150% Q_d and the velocity in the middle of the cross-section of the main canal is higher than at the sides (Figure 3-10). At the diversion point, deposition occurs in the inner bend, leading to the higher velocity at the middle and right side away from the branch in the case of 50% Q_d and the designed Q_d, respectively. When Q is increased to 150% Q_d, the velocity is distributed uniformly in the main canal. The increased velocity flushes the sediments away leading to less deposition in the upstream part of the canal. The sediments move further and accumulate near the diversion point, ultimately blocking the branch (Figure 3-8 right panel and Figure 3-10). Downstream of the diversion erosion occurs. In the case of 3000 ppm concentration, changes in discharge will not cause erosion but will reduce the deposition and will move the sediments further along the canal (for more details, see figure S 18 right panel in the supplementary document).

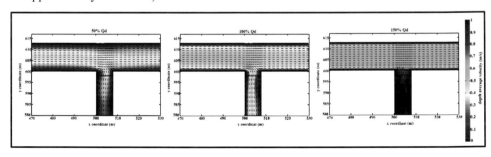

Figure 3-10 Effect of different discharges supplies (50%Qd, Qd, and 150%Q d) for non-cohesive sediments on flow velocity.

Variable water discharge has an impact on the deposition and erosion of both cohesive and non-cohesive sediments. However, the non-cohesive sediments are less dependent on shear stress than the cohesive sediments. Due to higher velocities, increasing the water supply by 150% leads to some erosion of already deposited non-cohesive sediments in the main canal, resulting in a lower bed level. In the branch canal, it reduces the deposition rate and increases the carrying capacity, consequently flushing the sediments to farmers' fields. Decreasing the water supply by 50% leads to more deposition in the main canal and a higher bed level (for more details, refer to the supplementary data see video S3A, video S3B, and video S3C which represent the effect of different water supplies on the velocity in the case of cohesive sediments entry. Video S4A, video S4B, and video S4C which represent the effect of different water supplies on the velocity in the case of non-cohesive sediments entry).

The differences in behaviour of cohesive and non-cohesive sediments under different flows are clearly visible in 2D/3D representation but would be missed by 1D longitudinal graphs.

3.4.3 Behaviour of cohesive and non-cohesive sediments under very high concentrations

This scenario shows the differences between cohesive and non-cohesive sediments under the very high concentrations of 3000 ppm and 1000 ppm respectively. The high concentration of cohesive sediments entering the irrigation system causes deposition along the main canal, especially near the diversion. Because of their small size, most sediment particles are carried to the end of the main canal, where a small deposition occurs downstream of the diversion, while the rest of the sediments are deposited in the branch canal. Due to the small sediment deposition in the main canal, the water depth is slightly reduced, leading to an increase in flow velocity. Without sediments, the flow velocity is 0.55 m/s from the beginning of the main canal till the diversion at 500 meters and 0.4 m/s thereafter. The flow velocity in the main canal increases to 0.7 m/s after the cohesive sediments enter the system.

Most non-cohesive sediments are deposited rapidly at the start of the main canal (Figure 3-11). Due to its weight and particle size, the resultant force acting on the sediment particles will hinder the sediments from moving downstream. The rapid deposition of non-cohesive sediments leads to substantially reduce water depths and cross-sectional areas, resulting in a substantial increase in velocity to 1 m/s (Figure 3-12), (for more details, refer to the supplementary data video S5A, video S5B and video S5C, where video S5A represent velocity when there is no sediment, video S5B represent velocity in the case of cohesive sediments entry and video S5C represent velocity in the case of non-cohesive sediments entry). Due to high sediment deposition before the diversion on the side close to the diversion, the velocity and carrying capacity are decreased so the deposition will occur and the sediment accumulation on this region will block the branch canal and no water can enter.

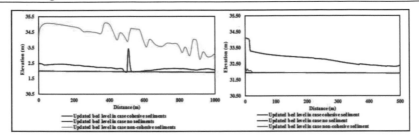

Figure 3-11 The change in canal bed levels for different sediment types.

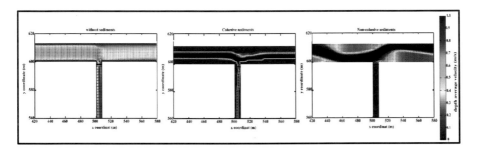

Figure 3-12 Velocity in the main and branch canals for different sediment types.

Different deposition patterns under the different scenarios in a 2D perspective are shown in Figure 3-13. The non-cohesive sediments are deposited in a meandering pattern, just like in braided rivers, creating 'sand bars' near the diversion and less deposition on the opposite side of the diversion. The blue colour indicates the negligible deposition on the canal bed. The higher deposition rate of the non-cohesive sediments leads to a higher bed elevation (as shown by the reddish and yellow colours in Figure 3-13). In the case of cohesive sediments, the deposition is higher at the sides of the canal than in the middle (as opposed to the meandering pattern of non-cohesive sediments), (for more details, see video S6A and video S6B in the supplementary data where video S6A represents the bed morphology development throughout a simulation period in case of cohesive sediment entrance and video S6B in the case of non-cohesive sediment entrance). The gate disturbs the flow to the branch canal, creating small eddies with spiral movement and leading to high sediment deposition. The large deposition of non-cohesive sediments near the diversion upstream of the branch canal blocks the branch, with no water flowing into it.

54

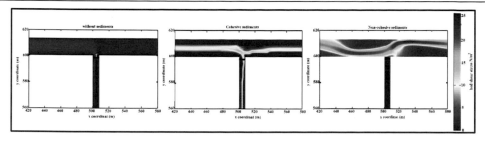

Figure 3-13 The differences in bed level between cohesive and non-cohesive sediments in the system.

Without sediments, the bed shear stress in the main canal is 2.2 N/m^2. It increases to about 12 and 22 N/m^2 in the presence of cohesive and non-cohesive sediments, respectively.

3.4.4 Interaction between non-cohesive and cohesive sediments

Most of the studies that simulate sediment transport in canals consider only one type of sediments, mostly non-cohesive, while in reality, rivers that supply water to irrigation systems contain a mix of cohesive and non-cohesive sediments. To understand the interaction between cohesive and non-cohesive sediments in canals, the authors compare the scenarios of pure non-cohesive sediments (30 ppm), pure cohesive sediments (100 ppm) and mixed sediments (30 and 100 ppm for non-cohesive and cohesive respectively).

In the case of pure non-cohesive sediments, a high deposition rapidly builds up in the upstream part of the main canal. Moving further, erosion occurs just before diversion and deposition just after the diversion. In the downstream of the main canal, erosion occurs (for more details, see video S7A, video S7B and video S7C in the supplementary data where video S7A represents the bed morphology development throughout a simulation period in case of cohesive sediment entrance, video S7B in the case of non-cohesive sediment entrance and video S7C in the case of mixed sediments). The cohesive sediments behave differently, with erosion occurring in the upstream part of the main canal (before the diversion) and deposition thereafter (Figure 3-14).

The deposition of non-cohesive sediments influences the flow characteristics which affect the transport and deposition of cohesive sediments and vice versa. The presence of cohesive (suspended) sediments slows the deposition rate of non-cohesive particles. This leads to a shift in deposition and erosion patterns as compared to pure sediment scenario. This effect is clearly shown in Figure 3-14, where the black line indicates the results of the mixed sediments scenario. The pink line shows the results of the scenario which considers that there is no interaction between the two types of sediments. The deviation between the pink line and the black line reveals the interaction between the two types of sediments.

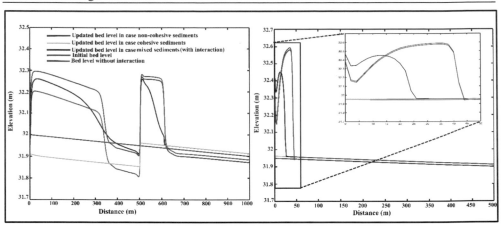

Figure 3-14 The indication of the interaction between the non-cohesive and cohesive sediments.

The pattern of mixed sediments more closely resembles that of the pure non-cohesive than the pure cohesive scenario, indicating that, although there is an interaction between the two types, the non-cohesive sediments are more dominant than the cohesive sediments, hence, the mixture resembles non-cohesive mostly. In the case of higher cohesive concentrations the mixture will resemble cohesive as shown in Figure S10, S11 in the supplementary file. In both cases, sediment deposition in the branch canal is very limited because most of the sediments have been deposited in the main canal, especially in the area close to the diversion location (Figure 15). These findings compare well with earlier studies on the effects of mixed sediments in rivers (Wu, 2016), concluding that in the case of a cohesive fraction lower than 10% of the mixture, then the mixture behaves as non-cohesive. As the cohesive fraction increases, the behaviour starts to resemble more cohesive (Wu, 2016).

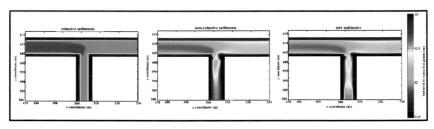

Figure 3-15 Bed level in the main and branch canal (with cohesive sediments, with non-cohesive sediments, with mixed sediments).

The velocity in the middle of the canal is higher than on the sides due to the friction with the sidewalls, (for more details, see video S8A, video S87B and video S8C in the supplementary data where video S8A represents velocity in case of cohesive sediment entrance, video S8B in the case of non-cohesive sediment entrance and video S8C in the case of mixed sediments).

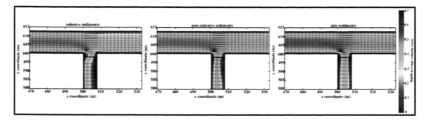

Figure 3-16 Velocity in the main and branch canal (with cohesive sediments, with non-cohesive sediments, with mixed sediments).

In Figures 3-15, 3-16 the differences are not significant because they show the average velocities over the vertical. Figure 3-14 presents the situation at the middle longitudinal line over the canal. In some irrigation systems after rainy season the deposits of cohesive sediments consolidate and are difficult to remove, so initially it would be cohesive sediment layer at the bed. For this reason the authors introduced an initial sediment layer of about 10 cm consisting of 50% cohesive sediments and 50% non-cohesive sediments In the case of one type of sediments, there is rapid deposition of non-cohesive sediments and erosion of cohesive sediments in the upstream part of the main canal. In the mixed sediments scenario, the rapid deposition of the non-cohesive sediment particles on top of the cohesive particles prevents the erosion of the cohesive sediment layer (Figure 3-17). Upstream of the diversion to the branch canal, there is an erosion of cohesive sediments because the non-cohesive deposition does not occur in this region. Downstream the diversion, there is a deposition for both kinds of sediments, while in the downstream end of the main canal there is only deposition of cohesive sediments (Figure 3-17).

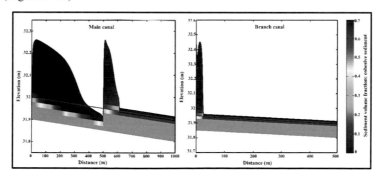

Figure 3-17 Sediment fraction distribution along the system.

While the interaction between cohesive and non-cohesive sediments are visible in 1D (Figure 3-17), the 2D representations produce a clearer representation of the phenomena (Extra 3D-Figures for all scenarios are provided in the supplementary data).

3.5 DISCUSSION

Cohesive and non-cohesive sediments behave very differently due to particle size, shape, weight and ionic charge. Because of the small size and chemical composition, cohesive sediments are carried in suspension and mostly deposited in the branch canals and fields. The coarse non-cohesive sediments are rapidly deposited in the upstream part of the main canal, while the branch canals and fields are almost free from non-cohesive sediments. The simulations show different sensitivities to main parameters such as concentration, discharge and particle size for cohesive and non-cohesive sediments (Table 3-3).

Table 3-3 Some parameters affecting on cohesive and non-cohesive sediments

Parameter	concentration	discharge	velocity	shear stress	d50
Cohesive sediments	(+)	(+++)	(NA)	(+++)	(NA)
Non-cohesive sediments	(++)	(++)	(++)	(NA)	(++)*

Where (+++) very sensitive, (++) sensitive, (+) weak, (NA) = not applicable

* Details in the supplementary document

Table 3-3 represents the sensitivity of cohesive and non-cohesive sediments to the change of key parameters in the tested scenarios. Scenario 1 ("changing sediment concentrations") shows that non-cohesive sediments are more sensitive to the change of concentration than cohesive sediments. On the other hand, scenario 2 ("changing discharge") shows that cohesive sediments are more sensitive to the change of water flow than non-cohesive sediments. However, the velocity (v) is not explicitly included in the equations governing the cohesive sediments. Hence, the sensitivity of cohesive sediments towards velocity cannot be determined (NA in Table 3-3). Similarly, the shear stress is not included in the equations for non-cohesive sediment transport and hence it is not applicable (NA). Additionally, the D_{50} of cohesive sediments is not included in the equations for cohesive sediment transport and hence it is not applicable (NA). This is a limitation of the equations used for cohesive sediments.

Based on the differences in sensitivity to the different parameters presented in Table 3-3, the deposition/erosion patterns in the system are very different between cohesive and non-cohesive as summarized in Table 3-4.

The use of 2D/3D models to show the differences between cohesive and non-cohesive sediments in irrigation canals

Table 3-4 Other differences between cohesive and non-cohesive sediments

Process	Deposition	Erosion	Time for deposition	Deposition rate	Maintenance	Mathematical simulations
Cohesive sediments	In the branch canal	upstream of the main canal	long	Low	long interval	1.5 days
Non-Cohesive sediments	upstream of the main canal	in the branch canal	short	High	short interval	3 days

The use of 2D/3D models was proven successful in simulating sediment transport and other 3D flow phenomena in rivers, estuaries and coastal areas (Gebrehiwot et al., 2015; Lesser et al., 2004; Van der Wegen et al., 2011). Furthermore, 3D models have been proposed to present sediment transport and flushing applications in sewer systems (Schaffner, 2008). However, because of their tremendous computational cost, they were not recommended for large-scale tests (Caviedes-Voullième et al., 2017). 2D models might be more suitable to represent and predict sediment transport in sewer systems, which is strongly determined by the channel geometry and the (probably 2D) velocity field (Caviedes-Voullième et al., 2017). In the current study, the 2D/3D modes of Delft3D perform well where they provide additional insights in representing the sedimentation/erosion patterns through and along the irrigation canals.

The Delft3D model was shown to perform well in several theoretical, laboratory, and real-life situations (Lesser et al., 2004). Despite the fundamental differences between rivers and canal systems, such as b/h ratio, sidewall friction, and the existence of weirs and gates, Delft3D has shown to perform satisfactorily in both rivers and irrigation canals (Theol, S. et al., 2019b). The structured mesh included in Delft3D can easily be stretched along the straight parts of the canal to achieve efficient simulation results, even when running in 3D mode.

This chapter uses Delft3D to illustrate the differences in behaviour and in erosion/deposition patterns between the cohesive and non-cohesive sediments. The simulations were run in 2D (running time or CPU time is one day and a half) and 3D (CPU time is 3 days), the graphs were presented in 1D for longitudinal canal sections, 2D for plan view and 3D for the cross-sectional area. The interpretation of model results using a 3D visualization is difficult due to the length-width-depth ratio of irrigation canals. Additionally, the high visual graphic output of 2D models makes it easier for engineers to convey their results and concepts to non-technical stakeholders.

In reality, most of the irrigation canals withdraw water contain mixed sediments (cohesive and non-cohesive). In Sudan, in the Gezira scheme, there is mixed sediments with 70% cohesive sediments and 30% non-cohesive sediments as stated by (Osman, 2015), however, Osman considered only the cohesive sediment. Osman et al. (2016) used the FSEDT model that simulate cohesive sediments in isolation, her model is able to predict the cohesive sediments only.

Other studies of sedimentation in irrigation canals use simulation tools designed for non-cohesive sediments such as the SETRIC model (Munir, 2011; Paudel, 2010). Due to the complex behaviour of sediment mixture, researchers assume the sediment as non-cohesive sediments only, on the other hand, these models were able to predict only the non-cohesive sediments. In this research, we aim to address this by introducing new method considering the mixed sediment (both cohesive and non-cohesive).

The fraction of cohesive sediments is substantial, this may lead to inaccurate results due to the interaction between cohesive and non-cohesive sediment particles. The presence of cohesive (suspended) sediments slows the deposition rate of non-cohesive particles. Simulations of river systems show that the particle interaction effect depends on the composition of the mix (Wu, 2016). When the cohesive sediments fraction is low, the mix behaves as non-cohesive. However, if cohesive fractions increased, the sedimentation patterns start resembling those of cohesive sediments (Wu, 2016). Also in irrigation canals when there is a higher cohesive sediment concentration, the mixture will resemble as cohesive sediments. The Delft3D model adequately captures both the behaviour of the two types of sediments and their interaction.

Despite using a simplified canal network, we believe the results are applicable to other canals with similar configuration and the findings have implications for more complex cases as well. However, there are other factors which are not addressed in this chapter such as discharge ratio, offtake location and angle, structure design, which influence sediment distribution. These were outside the scope of this chapter and would require more and different simulations.

3.6 CONCLUSION

Compared to the previous studies on sedimentation in irrigation canals in which 1D models were used, the use of Delft3D provided additional insights into the behaviour of cohesive and non-cohesive sediments. The deposition patterns are not uniform along the canals and are not evenly distributed in the canal cross-sections. In particular, near offtakes, diversions, and canal structures, the 2D/3D patterns are clearly visible. Knowing the different deposition patterns could help stakeholders in the planning of required canal maintenance. Based on this information, the stakeholders can decide which gate operation can be followed that minimize the undesirable deposition, also they will know when and where canals should be cleaned as a maintenance practice.

The authors adapted Delft3D for the use in irrigation canals but further work is needed to make the model fully compatible for the use in canals. In particular, running the Delft3D model in 3D mode takes a long time (3 days for a mix of sediments for the simplified canal layout). Running Delft3D for a more complicated canal network will be a challenge. The 2D mode cuts the running time by half and provides an adequate representation of sediment patterns, except near offtakes and structures.

The scenarios tested in this chapter clearly show the differences of deposition and erosion patterns between cohesive and non-cohesive sediments, regarding the location in the longitudinal direction and within the cross-section. The cohesive sediments are more sensitive to the changes in discharge and shear stress than non-cohesive sediments. On the other hand, non-cohesive sediments are more sensitive to the changes in concentrations than cohesive sediments. Lastly, the simulation in Delft3D clearly illustrates the interaction between cohesive and non-cohesive particles, in which the presence of cohesive sediments delays the deposition of non-cohesive sediments. This leads to depositing the non-cohesive sediments further downstream in the canals as compared to the scenario in which there are no cohesive sediments. This could have implications for the accuracy of previous studies in which non-cohesive and cohesive sediments are modelled in isolation, in particular where the irrigation system draws water from natural rivers.

DATA AVAILABILITY STATEMENT

All the dimensions of the model, the boundary condition and some criteria used in this article and the complete models for different scenarios used in this chapter are available in this link:

https://drive.google.com/open?id=1x4N7kL5Ee17QNyS6TDzo53XumnRb8uHg

SUPPLEMENTARY DATA

Supplementary document, videos S1-S8 and other supplementary data are available in: https://drive.google.com/open?id=1l8VTSR27BLNKqfFB2bMTcBCk4YNiVYch.

4

WHAT IS THE EFFECT OF GATE SELECTION ON THE NON-COHESIVE SEDIMENTATION IN IRRIGATION SCHEMES? – A CASE STUDY FROM NEPAL

This chapter has been submitted as:

Theol, S., & Bert Jagers, & J. Rai & F. Suryadi, and C. De Fraiture. What is the effect of gate selection on the non-cohesive sedimentation in an irrigation schemes? Submitted to the Water Resource Management journal.

What is the effect of gate selection on the non-cohesive sedimentation in irrigation schemes? –
A case study from Nepal

ABSTRACT

In order to cover the crop water requirements, the canals are operated to transfer the desired amount of water to the field canals by using the flow control structures like gates and weirs. In this chapter the impact of gate operation and selection of gates on the deposition of non-cohesive sediment is examined. The Delft3D model is used to simulate the effects of different scenarios regarding gate operation and the location of the gate that is opened. The model results showed that the gate selection affects not only hydraulic parameters but also morphological parameters. It was found that opening the gates closer to the offtake resulted in less sediment deposition at the entrance of the branch canal as compared to opening the gates further away. Gate selection can be used as tool in sediment management. By alternating the opening of different gates sediments which are already deposited by opening one gate can be eroded when another gate is operated, thus minimizing the additional cost of sediment removal. The use of Delft3D proved beneficial as the selection of different gates leads to asymmetric sediment deposition patterns which would be missed when using a 1D model.

4.1 INTRODUCTION

Water flows in irrigation canals tend to change depending on the amount of water demanded by crops or the supply of water in the irrigation system. This change in flow is assured by the operation of flow control structures such as gates and weirs. Canal operation results in unsteadiness of flow which is contrasting the steady and gradually uniform flow as is typically assumed in canal design (Depeweg et al., 2014). A minor variation in the flow characteristics may result in a major effect in the sediment transport patterns, leading erosion and/or sediment deposition in canals and around hydraulic structures in case of sediment-laden water. To keep a canal sediment-free, the sediment transport capacity of the canal should be maintained as foreseen in design criteria. However, heavy sediment load, badly designed canals and poor canal operation and management may result in undesirable erosion and sediment deposition. Canal operation is a crucial issue in sediment transport in irrigation systems, with gates being opened in varying heights and duration (Munir, 2011; Osman et al., 2016). This chapter investigates how different gate operation scenarios, including opening different gates and combinations of gates, can be used for minimizing undesirable sedimentation in irrigation canals, especially in the vicinity of flow control structures.

Different studies regarding sediment transport in irrigation canals showed the relationship between sediment problems and canal operation and management. (Depeweg & Paudel, 2003), employing different scenarios of canal operation, show that sediment problems in the Sunsari Morang Irrigation System in Nepal could be avoided or minimized by adapting gate operation plans. This finding was later confirmed by (Paudel, 2010). Observations by (Munir, 2011) confirmed that hydraulic efficiencies of the canals that are operated based on the supply had not affected by the sediment transport. Sediment depositions accumulated when canals run with low discharge can be flushed away during times of peak flow by regulating the operation (Munir, 2011). To reduce the deposition in the high concentrated sediment periods, Munir (2011) recommended operating the canals in supply based operation. From these studies, it is clear that canal operation and the method it is operated have an effect on sediment transport.

Previous studies were done using 1D models which simulate the patterns of the sediment erosion and deposition in the longitudinal direction of the canal, assuming that the sediments distribute just like water. However, because of unsteady and variable flow due to canal operation, sediment depositions are unequally distributed, especially near off-takes and structures. The effect of unequal distribution might be missed when using 1D models. To better representing the sedimentation patterns along and across irrigation canals, 2D/3D models are needed.

There are several mathematical 2D/3D models for the analysis of sediment behaviour in rivers, estuaries and lakes. Far fewer models were developed for simulating sediments in irrigation systems and most of them are limited because of being 1D and considering the equilibrium condition only. In this research, the well-established Delft3D model was selected because it can be run in 2D/3D model, simulating sediment transport for the non-equilibrium conditions in two dimensions. The other benefit of Delft3D is its ability to simulate the operation of gates with real-time control to show the morphological changes in canal beds.

This chapter will use the Delft3D model to analyze the impact of gate selection on the deposition and distribution within the cross section of non-cohesive sediment in the Sunsari Morang Irrigation system in Nepal. It builds on earlier work by (Yangkhurung, 2018) and (Paudel, 2010).

4.1.1 Delft3D

The Delft3D model is a multi-dimensional model developed by Deltares (Deltares, 2016). It can calculate non-steady flow and sediment transport phenomena in 2D/3D mode.

Delft3D model has been chosen to be used in this study for four reasons: 1) to represent the effects of gate selection to be operated on the non-cohesive sediments behaviour, 2) to show the settling basin effect on the deposition of the non-cohesive sediments, 3) to clarify the non-uniform flows around structures and offtakes, 4) to better simulate longitudinal and cross-sectional deposition patterns. The Delft3D model mostly has been used for rivers (Flokstra, 2006; Javernick et al., 2016; Parsapour-Moghaddam & Rennie, 2017) and for estuaries (Elias et al., 2001; Gebrehiwot et al., 2015; Lesser, 2009; Roelvink & Van Banning, 1995; Van der Wegen et al., 2011). Recently the model was also applied for irrigation canals (Theol et al., 2019). For the application in this chapter the real-time control (RTC) module is used which permits to change the gate height during the simulating period (Deltares, 2016). The technical details of the Delft3D model are described in chapter 1.7.

4.1.2 Study Area

One of the biggest schemes in Nepal is the Sunsari Morang Irrigation System (SMIS), which was constructed by the Government of India under the bilateral agreement between Nepal and India in 1954 and was handed over to Nepal only after two years of trial operation (Nippon Koei, 1995). The project aimed to irrigate 68,000 ha of Sunsari and Morang districts of Eastern Terai region of Nepal (Yangkhurung, 2018). The Koshi River, the source of the system, is highly sediment loaded (Paudel, 2010).The average monthly flow carried by the river is in the range of (500 to 6,000 m^3/s) while the annual flood discharge reaches to 7,000 m^3/s (Devkota et al., 2012). Various measures like a pre-settling basin (Figure 4-1 and 4-2) with a flushing arrangement and desilting basin with two dredgers have been applied to diminish the sediment problem in the system (Paudel, 2010).

Though the sediment problem has been reduced, the undesirable erosion and deposition are still evident from the raised canal bed, clogged structures and the high investment requirements to remove sediment from the system. This may be because, during the design of the system, the selected silt factor was small or the criteria of the sediment transport might have been ignored (Paudel, 2010). Furthermore, the canals in SMIS are unlined which makes the system more vulnerable to the sediment related problems.

The SMIS was designed as a supply-based system. However, during the modernization phase, the system changed to a demand-based system (Paudel, 2010). A rotation mode of 1:2 was followed for the water delivery schedule, where the offtakes are divided into two groups and providing the water for one group only per time (Paudel, 2010). However, due to the high demand in peak season and low flow in the source during the off-peak season, there was insufficient discharge in the Chatara Main Canal (CMC), leading to deviations in the developed cropping calendar. In order to supply a constant discharge, the canals are operated in rotation. The rotation duration for each canal depends on the water availability in the system. The water availability depends on the sediment concentration and on the river flow which in turn depends on the rainfall (DFID, 2006). In addition, most of the canals in the system do not have or follow canal operation plans. Canal and gate operation seem ad hoc and is not documented. Therefore, SMIS was selected for the study because of its vulnerability to the sediment problem and its potential to provide efficient and effective irrigation to a large agricultural area.

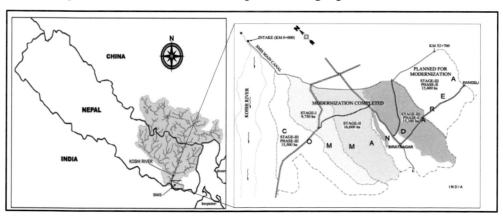

Figure 4-1 Location and layout of SMIS (Department of Irrigation, 2003)

What is the effect of gate selection on the non-cohesive sedimentation in irrigation schemes? –
A case study from Nepal

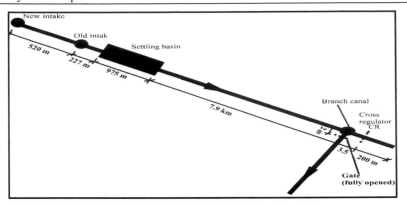

Figure 4-2 Case study schematization

4.2 METHODS

4.2.1 Data

For this study, the main canal and Sukhasaina secondary canal (S1) of SMIS were selected as this is the first major branch canal which includes proper water regulating structures. The S1 has its offtake at chainage 9.4 km of the main canal and irrigates about 8,146 ha with a discharge of 5.7 m³/s (Mishra, 2016). From Google earth, the authors took the canal layout while the canal geometry was taken from various documents and later verified with the staff of SMIS (Table 4-1).

Bathymetry

Table 4-1 Canal Geometry of the study area

	Main Canal	**Branch Canal**
Bed Width, b (m)	18	6
Canal Depth, H (m)	6	4
Side Slope, z	1.5	1.5
Canal Slope, S	0.00014	0.00014

Roughness

According to (Chow, 1959) the Manning's roughness (n) for an earthen canal, meandered with grass and some weeds is estimated at 0.03 s/m$^{1/3}$. This characterization is close to the existing canals in the system, and therefore the value of (*n*) for the main canal is taken as 0.03 s/m$^{1/3}$.

Sediment data

The sediments entering SMIS are mostly fine sediments with median diameter (d_{50}) of less than 200 µm (Paudel, 2010). The median diameter of 100 µm was used for representing the sediment size of the Sitagunj canal (S9) of SMIS by Paudel (2010) which is downstream of S1. For this reason, the median diameter of 100 µm was used to represent sediment size of the S1 canal, with a specific density of 2650 kg/m^3.

The concentration of sediment in the river changes according to the time of year. In the winter the concentration is less than 0.2 g/l while in monsoon season the concentrations are peaking to more than 3 g/l.

The gate just after the pre-settling basin is closed if the sediment concentration exceeds 0.3 g/l (Paudel et al., 2014). Therefore, the sediment concentration of 0.3 g/l with a median diameter of 100 µm is selected for the simulation.

Gates

For most offtakes, the water levels in CMC are usually controlled by cross regulators, while in the branch canal this is done by manually operated vertical steel gates (Mishra, 2016; Renault & Wahaj, 2006). The cross regulator (CR) in the main canal consists of a total of eight gates with 4 gate panels, each of 1.5m wide in 2 bays. The head regulator (HR) is similar but has 4 gates and two gate panels. For simplicity, in the simulation 4 gates and 2 gates are considered for CR and HR respectively, each with 3m wide.

4.2.2 Model setup

For validation purposes, the grid of this case was built based on the layout and the dimensions of the entire main canal and branch canal, including the settling basin. For the simulations of the scenarios, the results were analyzed zooming in to the section around the weir and outlet to the branch canal, excluding the settling basin. The bathymetry is developed based on the canal dimensions, the bed level of a known point, and canal bed slope. The discharge is the upstream boundary condition in the main canal which is kept constant during the simulation period. The Q-h relationship is the downstream boundary of the main canal. The flow resistance due to a barrier is dependent on the blocked flow by the gate, where the gate prevents the flow located in the top grid cells. Depending on the water elevation and the gate level, the number of blocked layers will be specified. The depth-averaged analysis showed that the energy loss coefficient (C_{loss}) is depending on the gate contraction, where it becomes zero when the gate is fully completely open. In the Delft3D model the appropriate energy loss coefficient (C_{loss}) for each gate must be specified (Deltares, 2016). However, the energy loss coefficient for the gates is not known; for this reason, it was assumed to be 0.9, the default value, for all the gates in the model (Yangkhurung, 2018). The Real-time control tool simulates the gate operation.

The difference between the sediments flowing in and out of the system specifies the changes in the canal morphology. This change is computed for each cell using the sediment transport formula of the non-cohesive sediments developed by Van Rijn et al. (2000). The method developed by Van Rijn (1993) is used to calculate the settling velocity depending on the suspended sediment size (Deltares, 2016).

The initial conditions are set as water level = 109 m + MSL with initial sediment concentration for each type of sediments is 0 kg/m^3. The initial sediment layer in the unlined section of the canal is assumed at 50 cm, whereas in the lined sections and near structures, the initial sediment layer is assumed zero since they are considered non-erodible.

The designed discharge is 51.3 m^3/s, however, it is observed that some parts of the canal are often overflowing during canal operation (Renault & Wahaj, 2006), implying that the canal capacity is actually smaller than that the design discharge. For this reason, a smaller discharge is used to avoid overflowing (Mishra, 2016). The discharge is assumed at 30 m^3/s.

The upstream boundary condition is the constant discharge of 30 m^3/s with a sediment concentration of 0.3 kg/m^3. The discharge, sediment particle size and concentration are kept constant throughout the simulation. The Q-h relation based on the canal geometry is taken as a downstream boundary condition at the end of the main canal since the inflow is changing due to the water extraction at the outlets. For the branch canal the boundary condition is the water level, fixed at 104.97 meters at the end of the branch canal. This is done because the branch canal is flowing into paddy rice fields with a wide undefined outflow.

4.2.3 Validation

Hydrodynamic validation

In order to ensure that the model is working properly and giving reasonable hydrodynamic results, we have separated the main canal and branch canal and compared simulation results with the boundary conditions for each model. The upstream boundary condition is the constant discharge; the downstream boundary condition for each model is the Q-H relationship. Table 4-2 compares the calculated and the simulated water heights (H) for the main and the branch canal for a range of discharges (Q).

Table 4-2 Calculated and Simulated water depth for specific discharges for Main and Branch Canals

Main Canal			Branch Canal		
Q [m^3/s]	Calculated H, (m)	Simulated H, (m) By Delft3D	Q [m^3/s]	Calculated H (m)	Simulated H, (m) By Delft3D
7.2	1	1	2.5	1	1
23.3	2	2	5.2	1.5	1.5
30	2.3	2.3	6.3	1.7	1.7
46.9	3	3	8.7	2	2

From Table 4-2, it can be seen that for the provided boundary conditions, the simulated and calculated water depths are equal. For this reason, the hydrodynamics of the main and the branch domain are considered reliable as basis for the next step in the simulation.

After the hydrodynamics verification of the separated domains, the main and the branch canals are combined by using the domain decomposition tool and the new combined domain is run. The model results are checked with field data and with the results from another hydrodynamic model, DUFLOW (Table 4-3). DUFLOW is a 1D program which can simulate unsteady flow. From the field data, for a certain discharge, the gauge reading of one point in CMC is known. The water level at this site was compared with the simulated water level.

Table 4-3 Water level obtained from various sources at an observation point

	Flow (m³/s)	Water Level (m)	% variation from design discharge
Designed discharge	51.3	2.98	
Duflow Model	51.3	2.97	0.0064
Delft3D Model	51.3	2.91	-2.18

The results show that the simulated water levels of both models are comparable. DUFLOW slightly overvalues the water level at the observation point while Delft3D has slightly underrated it. The DUFLOW results seem to be nearer to the actual value than Delft3D's results. However, the known observation point may have a different value of the water level since DUFLOW does not calculate the water level for a certain point; it calculates it for the whole section. The Delft3D model, on the other hand, calculates the water level at the center of each cell causing a staggered effect which increases the sectional area, resulting in a lower water level. Delft3D considers many parameters that are ignored by DUFLOW such as the side slopes and roughness. However, the results are both fairly close to the field data as shown in Table 4-3. From this, it can be said that the model is able to satisfactorily mimic the real situation from the hydrodynamic point of view.

Morphological validation

Based on the model setup and the actual field conditions, the morphological model was prepared.

The actual situation: the entire main canal including the settling basin

To validate the sediment deposition and erosion, the simulation results of the main canal including the settling basin (Figure 4-2), is compared with real field condition. The simulation shows how the sediments start to rapidly deposit in the upstream part of the main canal. This increases the bed level and leads to raising the water level upstream the main canal. Some sediments move forward and settle at the beginning of the settling basin raising the bed on the right side. After the settling basin, there is erosion because of the abrupt contraction in the canal.

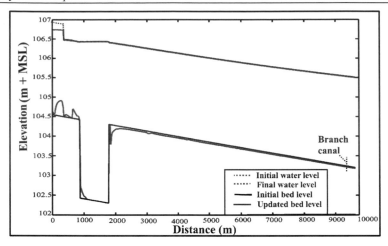

Figure 4-3 Bed level and water level updating in the main canal including the settling basin.

The velocity is reduced along the main canal especially in the settling basin where it reaches zero in the right bank, the reduction in velocity in this vicinity leads to sediment deposition (Figure 4-4). After the settling basin, the velocity reaches 0.6 m/s.

Figure 4-4 clearly illustrates how rapid deposition of the sediment can affect the water level. Also, it displays the role of the settling basin in trapping the non-cohesive sediment. The Figure 4-4 shows the results after one month. Running the same case for a longer period, leads to more sediments trapped and accumulated in the settling basin.

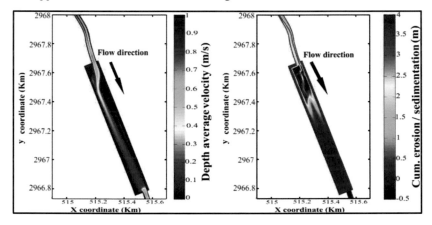

Figure 4-4 The relation between velocity and the accumulated sediment in the settling basin

Figure 4-4 presents a clear relation between the velocity and the accumulation of sediment. When velocity is decreased, the sediment deposition occurs. On the other hand, if the velocity increased, the deposition will be less.

Comparing the simulation results with actual situation as captured from Google Earth (Figure 4-5), shows the similarity of the sedimentation pattern in the settling basin, where the deposition occurs in the right side (as mentioned previously) of the settling basin.

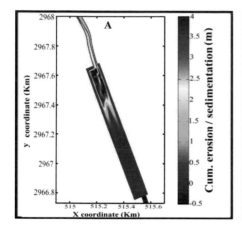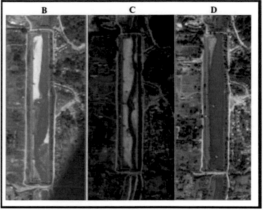

Figure 4-5 The similarity between Delft3D results and the actual situation captured from Google Earth (A: Delft3D results, B: the actual case in 2004, C: the actual case in 2005, D: the actual case in 2016.

From Figure 4-5, it can be concluded that Delft3D is able to satisfactorily represent the actual situation from a morphological point of view.

4.2.4 Scenarios

Different scenarios of the gate operation in the canal are tested using the Real-time Control (RTC) module. The RTC permits changing the status of the gate (opening partially, fully or even closing the gate) during the simulation period. This is done only for the gates of the Cross-Regulator (CR) which were operated while the gates of the Head Regulator are left fully open for all the scenarios (Figure 4-7). All CR gates are left fully open for half month, then the gates are fully opened or closed for a month, depending on the scenario (Table 4.4). Hydraulic input and sediment are maintained constant for all scenarios, changing only the opening of the gate. Scenario names are given as Scenario for Gate Operation (SGO) followed by a number. For the scenarios, the analysis is focused on the area within 1.15 km upstream of the offtake to the downstream end of the main canal (Figure 4-6), excluding the desilting basin.

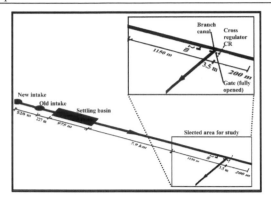

Figure 4-6 The selected area

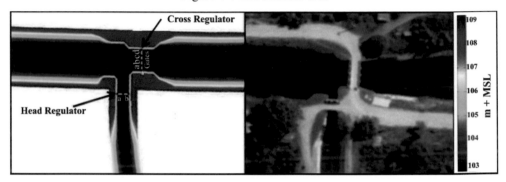

Figure 4-7 The location of Gates in the Canals in the model (left) and as seen in Google Earth (right)

Table 4-4 Gate operation scenarios

Simulation time		0.5 month	1 month		
Gates	Scenario name		Gate a (left bank)	Gate b (middle left)	Gate c (middle right)	Gate d (right bank)
All Gate Open	SGO1	All gates open	Open	Open	Open	Open
Two gates open	SGO2 (AB)		Open	Open	Closed	Closed
	SGO3 (BC)		Closed	Open	Open	Closed
	SGO4 (CD)		Closed	Closed	Open	Open
	SGO5 (AD)		Open	Closed	Closed	Open
One gate open	SGO6 (A)		Open	Closed	Closed	Closed
	SGO7 (B)		Closed	Open	Closed	Closed
	SGO8 (C)		Closed	Closed	Open	Closed
	DGO9 (D)		Closed	Closed	Closed	Open

4.3 RESULTS

4.3.1 Scenario 1 Reference case (SGO1) all gates are opened

For the reference case and the following scenarios, after excluding the settling basin (Figure 4-6), the model simulates a duration of 1.5 months. From Km 0 to Km 1.15, we see that the sediments start to deposit rapidly in the upstream and increased with time. This increase in the bed level again leads to raising the water level upstream of the main canal. The sediments move forward along the main canal due to the sufficient velocity and transport capacity to convey the sediments. The deposition of the sediments in km 1.15 is less than the upstream location, where it reaches to 0.4 m. After Km 1.15, the deposition increases because in this vicinity the water is diverted to the branch canal. Less water leads to less velocity, and hence an increase in the deposition. After the diversion to the branch canal, there is a contraction in the main canal, leading to increased velocity and reduced deposition in that location, and thereafter more deposition upstream the gates. Unlike the previous case, there is no erosion (Figure 4-8).

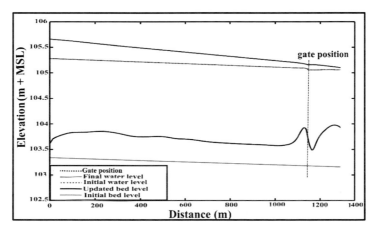

Figure 4-8 Bed level and water level updating in the main canal without the settling basin

The velocity is reducing along the main canal especially at the location of the diversion. This reduction in velocity leads to high sediment deposition (Figure 4-9). After the diversion, due to the contraction in the main canal at Km 1.17, the velocity increased and the sediment deposition decreased. After the contraction velocity decreased and sediment accumulation increased (Figure 4-9).

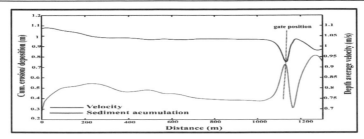

Figure 4-9 The relation between velocity and the accumulated sediment in the main canal.

Figure 4-9 presents a clear relationship between the velocity and the accumulation of sediment. When velocity decreases, the sediment deposition occurs and increases, when the velocity increases, the deposition will be lower.

4.3.2 Scenario 2 Gate Operation

In this scenario, the gates are operated during the simulation period by fully closing and fully opening for a certain time.

4.3.2.1 Operating two gates

4.3.2.1.1 Gate a and b (SGO2)

In this scenario the gates are opened for half month; then the gates are closed except the gates on the right side near the diversion (gates a & b in Figure 4-7) which stay open for one month. The sediment deposition in this scenario is higher than in the reference case (SGO1) from the beginning of the main canal till the location of gates, where the deposition is much higher (Figure 4-10) because of the reduction in velocity there. In this scenario compared to the reference scenario the bed level is higher the right side and lower at the left side of the canal (Figure 4-11). Downstream of the gates, the velocity higher because of closing two gates, causing erosion. Therefore, compared to the reference case, the bed level directly downstream of the gates is lower at the right and left sides and higher in the middle of the canal (Figure 4-11). In the branch canal, the deposition is almost the same as in the reference case, though a slightly less.

4.3.2.1.2 Gate b and c (SGO3)

Similar to the operation in scenario SGO2, the gates in this scenario are opened for half month; thereafter the gates are closed except the middle gates (b and c) which stay open for one month. The deposition in this scenario is also higher as compared to the reference case (SGO1). From the beginning of the main canal till the location of gates, the deposition is much higher (Figure 4-10) because of the reduction in velocity there. Similar locations of deposition, where the deposition upstream the gates on the right side is higher than in Scenario SGO1 and is lower on the left side, while downstream the gates the opposite occurs (Figure 4-11), the deposition in scenario SGO3 is less amount than in Scenario SGO2.

4.3.2.1.3 Gates c and d (SGO4)

In this scenario, the last two gates on the left side of the main canal far from the diversion side (c and d) are opened for one month after closing the other gates. The deposition in this scenario is lower on the right side than the deposition in the scenarios SGO1, SGO2 and SGO3, and higher at the left side for the whole main canal (Figure 4-11) except at the downstream the deposition is higher on both sides and erosion occurs in the middle due to the high velocity. The deposition is less because of the high amount of water transferred to the branch canal in this scenario. In the branch canal, the deposition is almost the same with the reference case a little bit less (Figure 4-10).

4.3.2.1.4 Gate a and d (SGO5)

In this scenario, the gates (a and d) stay open for one month after closing the other gates. The deposition in this scenario is higher than the reference case (SGO1) and SGO4 (Figure 4-10), and lower than in scenarios SGO2 and SGO3 from the beginning of the main canal till the location of gates, where the deposition is lower in both sides as compared to other scenarios and higher in the middle (Figure 4-11). Downstream the main canal, the deposition is higher than the reference case (SGO2) and SGO3 but lower than scenarios SGO1 and SGO4.

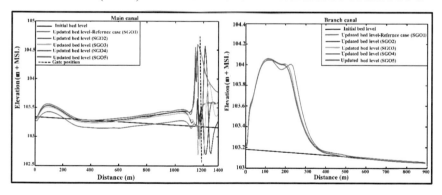

Figure 4-10 The bed level in scenarios with two gates opened

In these scenarios, opening only two gates from the total number of gates has an actual impact on the sediment transport, flow parameters, bed canal updating by the sediment deposition. Figure 4-10 shows that by using 1D representation, it is hard to tell the difference between the scenarios. While presenting in 2D/3D mode, we can clearly see the difference among these scenarios (Figure 4-11, 4-12) and see exactly the position of erosion and accumulation of sediments.

What is the effect of gate selection on the non-cohesive sedimentation in irrigation schemes? –
A case study from Nepal

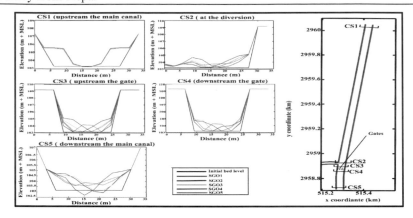

Figure 4-11 The bed level within the different cross-sections in scenarios with two gates
opened

The velocity has a major impact on the sediment deposition and erosion patterns along the
vicinity of the gate structure in the canal as discussed previously and shown in Figure 4-9. The
deposition is observable upstream the gate due to fewer velocity exists, while erosion was visible
downstream of the gate due to the higher velocity there. Scenario SGO1 by opening all gates
(free flow), the deposition occurs only on the canal sides due to the lower velocities. However;
the deposition was mainly concentrated near the gates that were closed during the simulation.

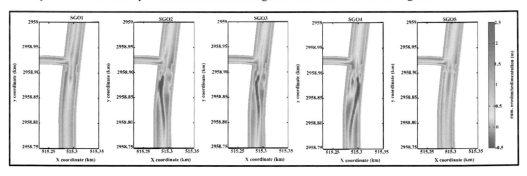

Figure 4-12 Cumulative erosion/sedimentation by opening two gates compared to the
reference case

In scenario SGO2, by opening the gates (a, b) there is erosion at the left side of the canal and in
the middle. In scenario SGO3, by opening the middle gates (b, c) erosion occurs in the middle of
the canal. Thereafter, in the main canal a significant deposition occurs, especially in the sides
where the velocity is low for all scenarios. In scenario SGO4, by opening the gates (c, d) there is
erosion on the right side and in the middle of the canal even there is an attempt to erode the sides
of the canal. While sediment deposited in the region near to the diversion due to lower velocities.
Furthermore, high sediment deposition before the offtake was observed in this scenario compared
to others. In scenario SGO5, by opening the gates (a, d) there is no erosion only deposition in the
middle occurs (Figure 4-12).

4.3.2.2 Operating one gate

4.3.2.2.1 Gate a (SGO6)

In this scenario, the gate (a) on the side of the diversion is opened for one month after closing all other gates. The deposition in this scenario is higher than the deposition in the reference case along the main canal except in the downstream of the gates until the end of the main canal (Figure 4-13). In the branch canal, the deposition is less than in the reference case, because, due to the closing of three gates in the main canal, a large amount of water is diverted to branch canal, leading to an increased transport capacity and reduced deposition from the start of the branch canal up to 400 m (Figure 4-13). After 400 meters to the end of the branch canal, erosion will occur due to the increase in the velocity. Within the cross-section the deposition downstream the gate and in the middle of the branch canal for case SGO6 is lower on both sides than the reference case, and is equal in the middle while downstream the branch canal the deposition is few millimeters higher on both sides than the reference case, small erosion in the middle of the cross-section (Figure 4-14).

4.3.2.2.2 Gate b (SGO7)

In this scenario, the gate (b) next to gate (a) will be opened for one month after closing all other gates. The deposition is similar as in scenario SGO6 in the main canal, branch canal and in the cross-sections (Figures 4-13 and 4-14).

4.3.2.2.3 Gate c (SGO8)

In this scenario, the gate (c) will be opened for one month after closing all other gates. The deposition is similar to SGO6 and SGO7 (Figures 4-13 and 4-14).

4.3.2.2.4 Gate d (SGO9)

In this scenario, the gate (d) on the other side away from the diversion side will be opened for one month after closing all other gates. The deposition is similar to SGO6, SGO7 and SGO8 (Figures 4-13 and 4-14).

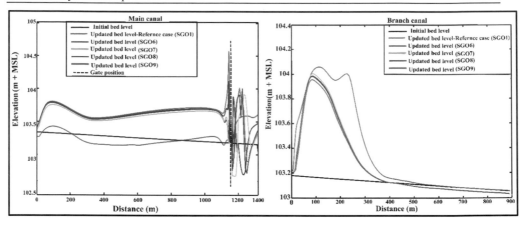

Figure 4-13 The bed level in scenarios with one gate opened

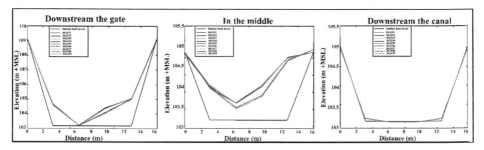

Figure 4-14 The bed level within cross-section in different scenarios

From Figure 4-14 it is clear that the effect of opening one gate on sedimentation is different from that of opening two gates. Where opening one gate provides less deposition downstream of the gate as compared to the deposition in the reference case and in the case of opening two gates. Also opening one gate produce less deposition in the middle of the branch canal on both sides than opening two gates and the reference case.

Opening gate (a) only leads to small sediment deposition at the upstream part of the branch canal as compared to scenario SGO7, SGO8, and SGO9. In the main canal downstream of the gate, the deposition occurs at the side far from the diversion and erosion occurs at the side close to the diversion due to high velocity. Thereafter, the erosion exists in the middle of the canal while deposition occurs in both sides of the bed due to low velocities (Figure 4-15). Opening gate (b) leads to erosion in the middle of the main canal, the erosion moves to the side far from the diversion. More deposition upstream the gate and then less deposition downstream the gate as compared to scenario SGO6 (Figure 4-15). Opening gate (c) in scenario SGO8 leads to more deposition at the upstream part of the branch canal and deposition upstream the gate. Downstream the gate, erosion exists in the side far from the diversion due to the high velocity and in the middle thereafter and deposition in both sides. Opening gate (d) has similar deposition and erosion patterns as the opening of gate (c), but more erosion.

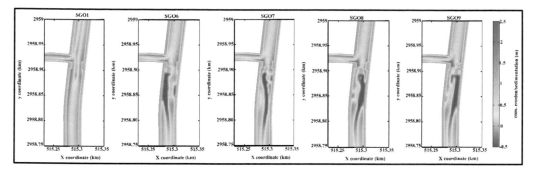

Figure 4-15 Cumulative erosion/sedimentation by opening one gate compared to the reference case

The results in Figure 4-14 and 4-15 show that using 2D/3D models is beneficial to clearly represent the difference in bed level within the cross-section under different scenarios. Knowing these differences can help decision-makers to choose the optimal canal operation by selecting the suitable gate to minimize deposition, while at the same time ensuring the required amount of water for crops is delivered. From these two scenarios, opening only one gate from the total number of gates has a greater influence on the sediment transport, flow criteria and the morphology updating for the canal bed than in case of opening 2 gates.

4.3.3 Other parameters

Water level

An important indicator in an irrigation system is the water level because it determines the amount of water diverted to the offtakes and fields. The scenarios with two gates closed (SGO2, SGO3, SGO4, and SGO5) have similar water level with small differences, but all are higher than the water level in reference case along the main canal till upstream the gate. While downstream the gates, the water level became lower than in the reference case.

Likewise, the scenarios with one gate closed (SGO6, SGO7, SGO8, and SGO9) exhibit a similar water level which is higher upstream of the gates as compared to the water level in reference case and also higher than the water level in scenarios with opening two gates. The higher water level tends to create a pool effect with decreased the velocity and accumulation of water to accommodate the flow into the branch canal. Downstream of the gate the water level is lower than in the reference case (Figure 4-16).

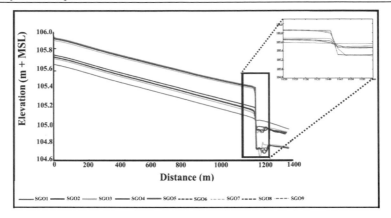

Figure 4-16 Water level along longitudinal mid-section of the main canal for different scenarios.

Velocity

For scenario SGO1, the velocity has a smooth transition of change along the main canal. While within the cross-section, the velocity in the middle is higher than the velocity in both sides (Figure 4-17). When the gates are closed, the velocity upstream of the gate will be decreased. While at the gate and after the gate, velocity will increase. In scenarios SGO2 and SGO4, the velocity is high in the side close to the diversion but differs in magnitude. In scenario SGO3, the velocity is high in the side far from the diversion. In scenario SGO5 the higher velocity is in the middle of the main canal.

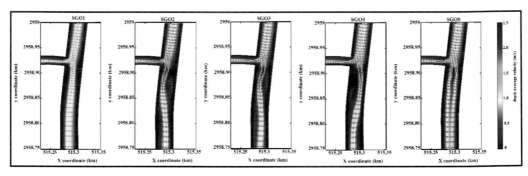

Figure 4-17 Depth-averaged velocity for scenarios with two gates opened

From Figure 4-17, an eddy can be seen downstream the gates for scenarios SGO2 and SGO3 in the side far from the diversion due to the disturbance caused by opening only two gates. In scenario SGO4, eddies are found at the side close to the diversion.

Opening one gate leads to less velocity upstream of the gates as compared to the velocity in the reference case. At the location and directly below the opened gate, velocity increases. Thereafter until downstream the main canal, the velocity is less than in the reference case. In scenarios SGO6 and SGO7, the velocity is high at the side close to the diversion but differs in magnitude. In scenario SGO8 and SGO9, the velocity is high in the side far from the diversion (Figure 4-18).

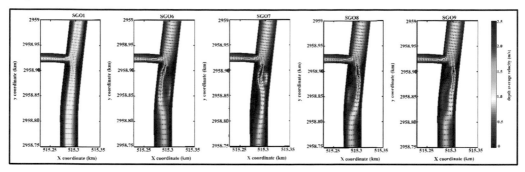

Figure 4-18 Depth-averaged velocity for scenarios with one gate opened

From Figure 4-18, an eddy can be seen downstream the gates for scenarios SGO6 and SGO7 in the side far from the diversion due to the disturbance caused by opening only one gate. In scenario SGO8 and SGO9, eddies are found at the side close to the diversion. Opening two gates has less velocity than the velocity in the reference case but higher than the velocity in scenarios with one gate open. The velocity was comparably lower at the sides of the canal and fairly minimal behind the closed gates. Based on Figures 4-17 and 4-18, these are the expected areas where the sediment deposition can occur due to the reduction in velocity, this confirms what we had seen in Figures 4-12 and 4-15 previously.

Sediment transport

In scenario SGO1, the sediment transport in the main canal upstream and downstream of the gates is almost similar, because of the slight change in the velocity. Unlike in other scenarios, the total sediment that is moved in the direction of the upstream of the gate is quite different from that at the downstream of the gate. The sediments in the upstream part are moved in the middle at a higher rate from both sides. The transport at the downstream is also affected by velocity: sediment moves where velocity is sufficient to convey them. Because the distribution of velocity within the cross-section is disrupted by the closure of the gates, the sediments being transported also change and are distributed unevenly within the cross-section. In the scenarios with two gates opened, the reduced velocity upstream the gates leads to reduced sediment transport capacity and high bedload occurs; most of the bed load settles which leads to having a more total load (Figure 4-19). Downstream of the gate the increased flow velocity due to the constriction of the gate closure, leads to an increased transport rate and reduced bed and total load. The increased velocity leads to erosion in the bed as well.

What is the effect of gate selection on the non-cohesive sedimentation in irrigation schemes? –
A case study from Nepal

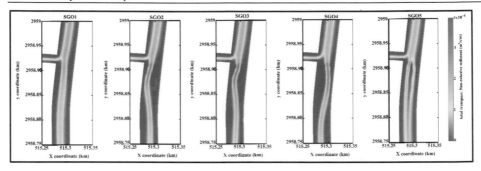

Figure 4-19 Total load transport in the main canal for different scenarios with two gates opened.

The accumulated sediments downstream of the gate on both sides of the canal are higher than in the upstream of the gate. This means when the sediment transport rate is high due to high velocity, less sediment deposition will occur and vice versa. Moreover, the sediment transport in the canal is dominated by a suspended load over bedload. The sill height of the gate blocks the bed load to transport further downstream. So, suspended sediments often flow toward the downstream of the gate and the bedload tends to settle at upstream of the offtake.

There is a difference between the volume outgoing sediments from the branch canal and the incoming sediment from the main canal into the branch canal. This means that the difference has been settled in the branch canal. The percentage of sediment eroded/deposited is calculated and presented in Table 4-5. However, any local erosion/sedimentation within these cross-sections cannot be adequately captured in this calculation due to the coarse resolution of the modelling.

In scenarios with one gate opened, the reduced velocity upstream of the gates lead to reduced sediment transport capacity and high bedload, leading to high total load; most of the bed load settles (Figure 4-20). While downstream the gate, there is no bedload because of the high transport rate due to the high velocity. Thereafter there is low bedload and low total load until the end of the main canal. The bedload in cases with two gates opened is less than the bedload in cases with one gate opened, because the transport rate is high due to the higher velocity.

84

What is the effect of gate selection on the non-cohesive sedimentation in irrigation schemes? –
A case study from Nepal

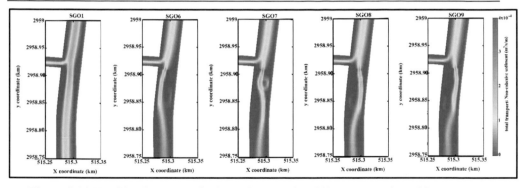

Figure 4-20 Total load transport in the main canal for different scenarios with one gate opened.

4.4 DISCUSSION

The basic function of the regulating structures is to convey the required amount of water to where it is needed, for example to the offtakes to the branch canals, and maintain a certain water level there. The reference scenario (SGO1) with all gates open diverts 29% of water to the branch canal. As expected after closing two or three gates in the regulator, more water enters the branch canal at the first time-steps.

Scenarios with two open gates divert 7% to 8% more water to the branch canal as compared to the reference case. Scenarios with one gate opened, divert about 19% to 21% more water to the branch canal as compared to the reference case (Table 4-5).

Table 4-5 Percentage of diverted discharge to the main and branch canals

% of Discharge Diverted	SGO1	SGO2	SGO3	SGO4	SGO5	SGO6	SGO7	SGO8	SGO9
Main Canal	71	64	64	63	63	51	52	52	50
Branch Canal	29	36	36	37	37	49	48	48	50

Scenario SGO5 diverts less water than scenario SGO2, SGO3, and similar SGO4 because in this scenario gates (a) and (d) are opened which are at the sides of the main canal where the flow velocity is less than in the middle because of wall friction. Opening gate b or gate c located in the middle of the main canal (scenarios SGO7 and SGO8 respectively) results in less water diverted to the branch canal than in scenarios SGO6 and SGO9.

Comparing the different gate operation scenarios with one or two gates open, it is clear that not only the number but also the location of the selected gate(s) has an impact on sediment deposition. Opening the gate which is on the side of the offtake leads to less sediment deposition at the upstream of the regulating gates (Table 4-6). Opening the farther gate leads to higher sediment deposition at the intake. This deposition will not only alter the hydrodynamic characteristics but also morphologic characteristics of the canal which can result in system inefficiencies as regulating gates are of crucial importance. The sediment depositions around structures are harder to remove and manage.

Table 4-6 Percentage of sediment diverted and stored at the main and branch canal for different scenarios

% of sediment	At the main canal								
	SGO1	SGO2	SGO3	SGO4	SGO5	SGO6	SGO7	SGO8	SGO9
Deposited along the main canal before the diversion	79	81	80	80	81	84	84	84	84
Diverted to branch canal	6	8	8	7	8	7	8	8	7
Stored at u/s of the gate	2	0	0	0	0	0	0	0	0
Deposited along the main canal d/s the diversion	14	12	12	13	11	9	8	8	9
Total %	100	100	100	100	100	100	100	100	100
	At the branch canal								
Deposited at branch canal	78	94	95	94	94	92	92	93	93
Transported out of the branch canal	22	6	5	6	6	8	8	7	7
Total %	100	100	100	100	100	100	100	100	100

The results in Table 4-6 show that when one gate is opened, the amount of deposited sediments in the upstream part of the main canal is higher than in scenarios with two gates opened and the reference case, because in the latter scenarios more water, and hence sediments, flow to the downstream of the main canal. Opening gates at the diversion side convert more water and sediments to downstream of main canal than when opening gates at the opposite side. Gates on the opposite side since their location away from the diversion side leads to having less sediment transport capacity and consequently higher deposition will occur and transported to the branch. The sediment stored at the branch canal is around 2% in the reference scenario while a negligible amount is stored in the other scenarios.

Among the scenarios with one gate opened, the location of the gate has a small impact on the amount of sediment diverted to the branch canal. In scenarios SGO6 and SGO9 the sediment is slightly less compared to other scenarios. Although the difference is small, it is worth mentioning that gate operation does have an impact on the sedimentation and erosion of the branch canal.

Additionally, in scenarios of two gates open, the amount of water transferred to the branch canal in the case of SGO4 was more than the reference case, SGO2 and SGO3 and similar to SGO5 as shown in Table 4-5. On the other hand, the amount of sediments transferred to the branch canal in the case of SGO4 was less than other scenarios of two gates open as shown in Table 4-6.

While in scenarios of one gate open, the amount of water transferred to the branch canal in the case of SGO9 was more than other scenarios as shown in Table 4-5. On the other hand, the amount of sediments transferred to the branch canal in the case of SGO9 was less than SGO7, SGO8 and similar to SGO6 as shown in Table 4-6.

In practice, usually, it is preferred that the gate operation to have less effect on the canal shape, in this case, the operation of the gate (SGO4) seems a better choice among the scenarios with two gates opened. The operation of the gate (SGO9) seems a better choice among the scenarios with one gate opened. It has ensured delivering a higher amount of water to the branch canal with fewer sediments.

A slight change in hydrodynamic characteristics changes the sediment transport mechanism and thus the canal morphology which in turn changes the water flow features in the canal (Munir, 2011). The results show that because of gate operation the change in the flow characteristics alters the amount of sediment passing through each cross-section different for each scenario. This means that gate operation has a significant effect on the hydrodynamic as well as the morphologic parameters in an irrigation system (Munir, 2011). Gate operation can be used as not only the as diverting the water but also as a sediment management technique. The incoming flow through time can erode the accumulated sediment and flushed them away. This is because of the change in the canal properties and the hydraulics of the canal. For this reason, it can be said that there is constant sediment erosion and deposition until the equilibrium state is reached, in which this condition can occur only in the case of non-cohesive sediments.

Operating the gate at the side of the offtake can minimize the sediment deposition at the entrance of the offtake, however, if we open the same gates regularly, then the canal geometry will be permanently altered due to the deposition and erosion resulted from this gate operation which is not good for equitable water distribution and for the proper system functioning. Additionally, it has been noticed that the deposited sediments during one gate operation can be removed during the other (Yangkhurung, 2018).

The results from comparison of the scenarios clearly show that the number of gates opened as well as their location (a, b, c or d) has an impact on sedimentation and erosion patterns in the both the main and branch canal. Not only the amount of sediment deposition but also the location in the cross section of the canal depends on the gate selection. Therefore, a canal operation plan along with the gate operation schedules can help to increase the canal efficiency and to reduce sediment removal costs. Alternating the gates to be opened during the irrigation period can help to reduce the sediments around the flow control structures, while at the same time delivering sufficient water to the branch canals. Sediment depositions can be eroded with the help of the canal operation itself, by alternating the gate to be opened, without investing extra money and labour for cleaning process.

This chapter considers non-cohesive sediments only. In reality, many irrigation systems take water from rivers which contain a mix of cohesive and non-cohesive sediments. In the next chapter we consider this.

4.5 CONCLUSIONS

Gate operation and gate selection have a significant impact on the hydrodynamic and morphologic parameters in irrigation canals. Opening the gate on the side of offtake resulted in less deposition of sediment at the entrance of the offtake than in the scenario where the gate on the other side of offtake was opened, which indicates that the selection of gates to be opened for delivering a certain amount of water had a major impact in the sediment erosion and deposition patterns.

Furthermore, the impact of gate selection was not only on the main canal, but it also extended to the branch canal. The deposition was clearly seen before and after the closed gates. The erosion and deposition pattern differs and has an asymmetric distribution along the canal and within the cross-section.

The simulation of sediment patterns under different gate selection scenarios is best represented by 2D or 3D models. Most previous simulation studies regarding sediment in irrigation canals use 1D models. While these models provide insight in the quantity of sediments in the longitudinal direction of irrigation canals, asymmetric deposition patterns will be missed. The scenario results tested in this chapter show that gate selection has a major impact on where sediments are deposited in the cross section. These effects would be missed by using a 1D model. The simulations in this chapter were run both in 2D and 3D mode. While the 3D mode provide better insights in the exact location of deposition/erosion, running times are prohibitly long for irrigation canal networks. Compared to 1D models, the use of Delft3D run in 2D mode is beneficial in visualizing the hydraulic and morphological aspects and providing insights in the sediment distribution along the canal and within the cross-section, while avoiding the long simulation times of 3D models.

5

THE ROLE OF GATE OPERATION IN REDUCING PROBLEMS WITH COHESIVE AND NON-COHESIVE SEDIMENTS IN IRRIGATION CANALS

This chapter is based on:

Theol, Shaimaa, & Bert Jagers, & F. Suryadi, and Charlotte De Fraiture. The role of gate operation in reducing problems with cohesive and non-cohesive sediments in irrigation canals. *Water* **2019**, *11*, 2572; doi: 10.3390/w11122572.

The role of gate operation in reducing problems with cohesive and non-cohesive sediments in irrigation canals

Abstract

Sediments cause serious problems in irrigation systems, adversely affecting canal performance, driving up maintenance costs and, in extreme cases, threatening system sustainability. Multiple studies were done on the deposition of non-cohesive sediment and implications for canal design, the use of canal operation in handling sedimentation problems is relatively under-studied, particularly for cohesive sediments. In this chapter, several scenarios regarding weirs and gate operation were tested, using the Delft3D model, applied to a case study from the Gezira scheme in Sudan. Findings show that weirs play a modest role in sedimentation patterns, where their location influences their effectiveness. On the contrary, gate operation plays a significant role in sedimentation patterns. Reduced gate openings may cause canal blockage while intermittently fully opening and closing of gates can reduce sediment deposition in the canal by 54% even under conditions of heavy sediment load. Proper location of weirs and proper adjusting of the branch canal's gate can substantially reduce sedimentation problems while ensuring sufficient water delivery to crops. The use of 2D/3D models provides useful insights into spatial and temporal patterns of deposition and erosion but has challenges related to running time imposing a rather coarse modelling resolution to keep running times acceptable.

5.1 INTRODUCTION

Improved irrigation water management plays a crucial role in enhancing crop production for food security. Sediment control in irrigation systems is of great concern for irrigation managers and farmers because sedimentation in canals and near structures often contributes to water management problems. Further, problems of heavy sedimentation loads may jeopardize the sustainability of irrigation systems due to the high costs of cleaning canals (Osman, 2015). Therefore, understanding the mechanisms underlying sediment transport in irrigation canals received substantial scholarly attention (Belaud & Baume, 2002; Jian, 2008; Jinchi et al., 1993; Mendez, 1998; Nawazbhutta et al., 1996; Paudel, 2010). However, most of these studies focus on system design and relatively few take into consideration the effects of irrigation structures and the operation of gates.

Crop water requirements are not constant but change throughout the season depending on the crop growth stage. Consequently, flows in canals that supply water to fields are variable depending on the use of control structures such as gates and weirs. Structures often cause unsteady flow in the canals, even where they are designed for steady or uniform flow. The change in flow affects the sediment transport which leads to sediment deposition and erosion in different locations of the canal. Even though canals are typically designed to keep the bed free from sediments and convey sediments to fields, the improper placement and operation of gates and weirs in the absence of optimal canal operation plans may lead to deposition and erosion of sediment in canals and reduce canal performance. The impact of canal operation on sedimentation problems in irrigation systems deserves more attention in modelling studies of irrigation systems.

Examples of studies simulating the effect of canal operation on sediment transport include Depeweg and Paudel (2003) in the Sunsari Morang system in Nepal and Munir (2011) in the Machai Maira Branch Canals in Pakistan. However, both studies only considered non-cohesive sediment, mostly transported as bed material. In reality, many irrigation systems deal with a mix of coarse (non-cohesive) and fine (cohesive) sediment. Dealing with sedimentation in irrigation canals becomes more complex in case of cohesive sediments due to its physiochemical properties and inter-particle forces. Most studies regarding cohesive sediment behaviour have been done in rivers and estuaries (Celik & Rodi, 1988; Guan et al., 1998; Liu et al., 2002; Lopes et al., 2006; Van Rijn et al., 1990; Wu et al., 1999). There is great need to study the mechanism of cohesive sediment transport in irrigation canals (Theol et al., 2019), in particular under different scenarios of gate operation.

The impact of gate operation on the cohesive sediment in the Gezira Scheme in Sudan has been studied by Osman et al. (2017). They considered two gate operation scenarios: (1) a system in which water allocation is based on water duty and the cropped area and water is given by a fixed discharge for one week. This so-called indent system has been followed for several ago in Gezira system and (2) a system in which water supply is reduced based on the crop water requirement when sediment concentrations reach its peak. They found that the latter scenario performs best, reducing sediment deposition to 48%, primarily because the intake of the amount of sediment-laden water is reduced. Osman et al. (2017) and Munir (2011) both used a 1D model while the behavior of cohesive sediments is best reflected in 2D/3D models (Theol et al., forthcoming).

The main objective of this chapter is to investigate the role of gate operation in reducing the amount of cohesive and non-cohesive sediment in the canals using a 2D/3D model. This chapter builds on the work by Osman (2015) on the sediment deposition patterns in the Gezira irrigation scheme and uses the baseline data collected by her. However, we use a mix of cohesive and non-cohesive sediment and use Delft3D, a model that can be used in 2D or 3D mode (Lesser et al., 2004; Theol et al., forthcoming), to test different scenarios of weir height and duration of gate openings. We consider the Gezira irrigation scheme in Sudan as illustrative for many irrigation systems in semi-arid areas suffering from high sediment loads originating from river intakes.

5.2 MATERIALS AND METHODS

5.2.1 Model Selection

Using 1D models to study hydrodynamics in irrigation canals computationally is efficient, however, these models may not be representative in morphologic simulations. 1D-models have a simple ability to present several basic phenomena exist in nature which is usually found in three-dimensional (Lesser, 2009; Morianou et al., 2016). On other hand, 2D or 3D models can detect sediment movement and sediments patterns near offtakes and structures in more detail and simulate deposition and/or erosion locations within the cross-section in addition to those in the longitudinal direction. While beneficial from a morphological point of view, the biggest constraint of 2D and 3D models is the long simulation time.

To explain why we selected Delft3D, we compare three well-known 2D/3D models that that are able to simulate sediment transport in canals (Table 5-1), namely Delft3D, Telemac (Villaret et al., 2013) and Mike21 (Morianou et al., 2016).

Table 5-1 Comparison between different models

Features Available	Delft3D	Telemac	Mike 21
Grid construction	Structured with DD*		
	Unstructured (FM)**	FM	FM
Simulating: Cohesive sediments	Yes	No	Yes
Simulating: Non-cohesive sediments	Yes	Yes	Yes
Open-source	Yes	Yes	No
RTC***	Yes	No	Yes

*DD: domain decomposition, **FM: flexible mesh, ***RTC: real-time control

Structured grids with rectangular cells and areas are computationally efficient if aligned with long straight canals. However, in reality a canal system consists of main canals and branches, with large 'empty' areas in between (Figure 5-1). These 'empty' inactive parts, which fall outside the area of interest, render the structured grids inefficient since the model domain includes large inactive parts taking up unnecessary computation time. Unstructured grids (or flexible mesh) can model irregular shapes that only include the active parts of the channel networks. However, these grids mostly consist of triangular cells which are not conducive for long canals, since they cannot be stretched in stream direction, leading to a higher number of triangular cells and hence longer simulation time. One possible solution is to use an unstructured grid with quadrilateral cells aligned with the flow direction along the canal (e.g. Delft3D FM or Mike FM). A more efficient solution is to use a structured grid with the domain decomposition (DD) tool available in Delft3D. This tool allows to divide the grid in separate parts that can be modelled and compiled. In this way inactive parts can be excluded, substantially reducing simulation time. The latter method, combining structured grids and domain decomposition, was used in this chapter.

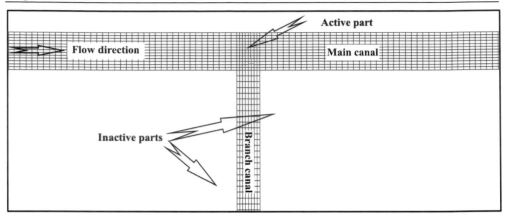

Figure 5-1 The active and inactive parts in the computational domain

The Real Time Control (RTC) tool enables changing weirs and gate settings during the simulation. This property can be activated in the Delft3D-FLOW input file by using the Rtcmod keyword. It allows simulating canal operation in which gates are opened and closed multiple times during the irrigation season. The morphological factor (Morfac or MF) feature in Delft3D further shortens the model running time and enables predictions of the morphologic developments in the medium term (months or seasons).

Comparing the three models Delft3D has all features necessary to simulate the effect of gate operation under scenarios of cohesive and non-cohesive sediment and their interaction. It is also open source and can handle non-steady flows.

Delft3D has been validated by (Lesser, 2009) for a series of simplified theoretical, laboratory and full-scale test cases. Furthermore, it was also validated against the results of prototype-scale measurements. A big advantage of numerical simulations is that there is no need to apply scale factors (Lesser, 2009), unlike physical morphological models where sediment scaling is a major problem. Numerical morphological models can be tested directly against both the laboratory observations and prototype-scale observations.

So far the Delft3D model has been used primarily for rivers (De Jong, 2005; Flokstra et al., 2003; Gebrehiwot et al., 2015; Kemp, 2010) and for estuaries (Lesser, 2009; Van der Wegen et al., 2011). The model has been used by Theol et al. (2019) for irrigation canals in both 2D and 3D modes. Running the model in 2D mode is to ensure better representation of the sediment processes and the large scale behavior with an acceptable simulation time period. Running the model in 3D mode provides information about the vertical and gives more details near structures.

5.2.2 Case study

The Gezira Scheme is the largest irrigation scheme in Sudan, serving 880,000 ha and taking water from the Blue Nile River which carries large amounts of sediment. Since its construction in 1920, the scheme suffers from sediment accumulation in the canals, representing a big challenge for the operation and maintenance. The annual costs of desilting were estimated at around US$12 million (Gismalla, 2009). The irrigation system consists of a network of main, major, minor and field canals. Two canals were selected for this study: the Zananda Major canal and Toman Minor canal, fed by the Zananda Canal. The Zananda canal is the first canal that takes water from the Gezira Main Canal by gravity irrigation (Osman, 2015) (Figures 5-2 and 5-3).

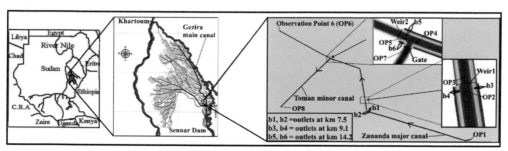

Figure 5-2 Scheme of the Zananda Major Canal and Toman minor canal.

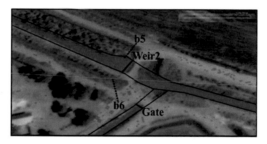

Figure 5-3 Location of Toman minor canal (Google Earth).

The location of the off-take is 14°01′42″ N and 33°32′33″ E. The Zananda canal is 17 km long and provides water to seven minor canals in which irrigate about 8,520 ha, one of these minor canals is Toman Minor canal. In Figure 2 the other minor canals are presented as outlets named b1, b2, b3, b4, b5, and b6. In the selected area, 75% of the sediment is silt and clay with grain sizes less than 0.063 mm (considered as cohesive sediment); the remaining 25% is fine sand as mentioned in the analysis of the bed materials done by (Osman, 2015). Osman et al. (2016) concluded that sediment is transported in suspension, based on sediment analysis. More details regarding the canal geometry and hydraulic parameters are presented in Table 5-2.

Table 5-2 Geometric data

Criteria	Major canal	Minor canal	Unit
Canal length	17	6.5	Km
Canal width	4 m from (0-9.1) km 3 m from (9.1-17) km	2	M
Canal bank height	5	4	M
Roughness (n)	0.029	0.029	$s/m^{1/3}$
Slope	0.0001	0.00005	-
Side slope	1:2 from (0-14.2) km 1:1 from (14.2-17) km	1:1.5	-
Structures	Weir 1 and weir2 with a height of 0.3 m, length of 3 m.	Gate fully opened	M

5.2.3 Model setup

Grid construction, bathymetry, and other parameters assumptions

We constructed a grid for the Zananda major canal of 17 km long and 4 m wide from the inlet till the first contraction after the first weir where the width becomes 3 m till the end of the major canal. The grid for the Toman minor canal is 6.5 km long and 2 m wide with eight observation points as depicted in Figure 1. We followed the grid quality criteria of Delft3D with the orthogonality = 0.05 (i.e. cells are almost perpendicular to each other which proved the most suitable grid setting for reducing the Courant number that causes model instability in the course of the simulation) and smoothness = 1.2 for both M and N directions. The grid for the major canal contains 1125 and 14 cells in the M and N-direction respectively. The grid for the minor canal is 581 by 6 cells. The grid size for the long straight canal is 18 m. The more accurate mesh size of about 1 meter is used in important areas such as near the structure and near the minor canal. To reduce the computation time, the network domain is divided into the major grid domain and minor grid domain. The simulation results for both domains are compiled using the Domain Decomposition tool (DD), which reduces the simulation time to 40% from the total simulation's time. Using the field data presented by (Osman, 2015), we took the elevation of the upstream part of the Zananda major canal as a starting point and built the bathymetry of the remainder of the canal based on slope, length, and canal geometry such as bed width, side slopes, and roughness. We estimate the design discharge of the major canal as 5.5 m³/s, based on available data and field reports (Osman, 2015). When canals are free of sediment, the flow is assumed to be as steady non-uniform flow during the time step (i.e., flow rates of the outlets do not change with time but depth of water varies with the location along the canal). For hydrodynamic reasons, the model is first executed without sediment to get a steady-state flow condition and check some crucial flow parameters such as velocity, water levels and the bed shear stress which is important in calculating the sedimentation and erosion of cohesive sediments. The steady-state flow condition was validated with results from the DUFLOW model following the method described by Theol et al. (2019). Due to the absence of the detailed field data regarding velocity and bed shear stress, we compare Delft3D results of water level to the DUFLOW model, where the DUFLOW model was previously calibrated by Osman (2015) against field data. Our water levels match those of DUFLOW within 5 cm. Osman (2015) validated the DUFLOW model against field data, the water level of DUFLOW match the field data within 3 cm.

After adjusting uniform bed roughness and wall roughness for hydrodynamic parameters, we test the scenarios assuming the entrance of sediment at a constant rate, evaluating the morphological changes in the canal bed after a simulation time of three months and comparing the results to the initial bed levels.

Model runs

The model was run for a simulation time of three months using a time-step of 0.6 seconds and a morphological factor (MF) of 10 using both 2D and 3D modes. The results of the 2D and 3D simulations look identical. In this chapter, the graphs are based on the 3D simulations. The small-time step is chosen to avoid the Courant number exceeding 1.0 which would destabilize the model. The MF enables the computation of the morphodynamics together with the hydrodynamics. This MF was used to speed up the changes of bed morphology by 10 times per time step, which reduces the time by a factor of 10. Thus, simulating the effective morphological changes over 3 months requires only a simulation period of 9 days. The MF approach simplifies the model setup and operation in comparison with other approaches (Li, 2010; Roelvink et al., 1998), in this way, the Delft3D model was capable to predict the changes in canal morphology over a long time span within small simulation time.

Two different computers were used in this study, One has simple specifications (dual-core Hp ProBook 6570b) and the other is a higher-performance computer is (quad core Hp Z Book15 G3); the latter reduced the simulation time by 40%. The CPU time was 3.5 days and 2days for 3D and 2D modelling respectively.

In this study the maximum concentration is assumed to be (C_b^l = 3 kg/m³ or 3000 ppm) for cohesive sediments, this concentration lies in the range of typical concentrations which are relevant for the Gezira Scheme (Osman, 2015). As input data in Delft3D, the settling velocity (W_s) is set to 0.12 mm/s which corresponds to the Krone (1962) formula for the aforementioned concentration. The value of the critical shear stress for erosion ($\tau_{cr,e}^l$) is set to 1 N/m². For the erosion parameter M¹ the default value of 0.0001 kg m⁻²s⁻¹ is used. For the critical shear stress for deposition, the authors used $t_{cr, d}$ = 1000 N/m².

The initial conditions are set as follows: water level = 34 m + (MSL) Mean Sea Level. The initial sediment concentration for each type of sediment equal to 0 kg/m^3, the canal bed is erodible (movable) limited by the available amount of sediment. The initial sediment layer assumed to be 20 cm consist of 50% sandy material and 50% muddy material. The boundary conditions are: discharge equals 5.5 m^3/s with sediment concentration is 3000 ppm and 100 ppm for cohesive and non-cohesive, respectively, as un upstream boundary condition. The downstream boundary condition for each canal was taken as Q-h relation which is based on the canal characteristics. For the other branches b1, b2, b6, they have been considered as only outflow, where each branch drags 0.5 m^3/s of water from the major canal.

Other parameter values regarding non-cohesive sediment are D_{50} = 100 μm (fine sand) with a specific density of 2650 kg/m^3. For the transport of non-cohesive sediment, we use the Van Rijn formula (Van Rijn, 1993).

Morphological comparison

Osman (2015) collected sediment data in the Gezira irrigation system in Sudan for the years 2011 and 2012. Given the difficulty of getting actual field data, we used the data from (Osman, 2015) to validate the Delft3D model.

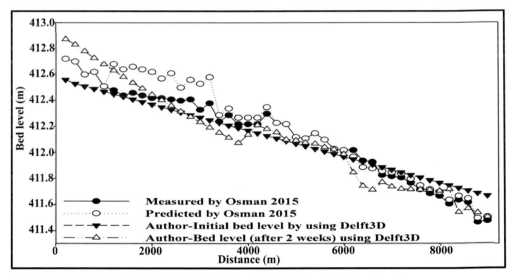

Figure 5-4 Model validation comparing results from the Delft3D model with field data collected by (Osman, 2015).

Figure 5-4 shows that the results obtained from Delft3D are qualitatively comparable with the field data measured by (Osman, 2015), giving confidence that the Delft3D model is able to replicate field conditions. Delft3D results' are also qualitatively comparable with the simulation results by (Osman, 2015). Observed differences in bed level are partly explained by differing modelling assumptions, the two simulation models use a different numerical technique. Osman (2015) used a series of quasi-steady-state computations in her model, whereas Delft3D a dynamic model showing changes in time for flow characters like velocity, water depth and bed level changes due to sediment transport over time.

5.2.4 Scenarios

To assess the effect of operation and structures, we tested different scenarios regarding gate opening and height of weirs; weirs and gates influence the Delft3D computation by changing the through-flow area (2D) and (partially) blocking specific layers in the 3D model. The structures do not directly affect the sediment transport; the sediment transports are affected indirectly by the changed flow patterns. Gate opening and weir heights vary according to the scenario (Table 5-3) while sediment concentration and other parameters are kept constant during the simulation. Regarding the gate operation, the Real-time Control (RTC) module is applied because this tool allows us to open the gate fully, partially or fully close during the simulation.

Table 5-3 Scenarios in the study.

Scenario	Description	Remarks
1. Reference case	Full open gate and fixed weirs' heights.	Gate fully opened; w1, w2 with fixed height at 0.3m.
2. Effect of the weirs	a. Setting of the upstream weir height b. Setting of the downstream weir height c. Setting of both weirs.	Gate fully opened; lowering or removing, raising the weir (0 m, 0.6m).
3. Effect of the gate	a. gate setting with constant openings	Lowering the gate (0.2m–0.8m); weir1 and weir2 with fixed height at 0.3m.
	b. gate setting with variable openings	Operation plans for the gate; weir1 and weir2 with fixed height at 0.3m.

5.3 RESULTS

5.3.1 Reference scenario

In the reference case, the gate in the Toman minor canal is fully opened while in the Zananda major canal, the height of both weirs is fixed at 0.3 m. During the simulation, sediments start depositing in the upstream part of the major canal (Figure 5-5). The cohesive sediment deposit gradually, distributed over the major canal while the non-cohesive sediments deposit mostly in the upstream of the canal. Because of the mixed sediment and interaction between cohesive and non-cohesive particles, the sedimentation pattern differs from the case of pure non-cohesive sediment. In the case of pure non-cohesive sediment, the heavy non-cohesive particles would rapidly deposit in the upstream the major canal. In the case of mixed sediment, some non-cohesive sediments are transported all the way to the downstream of the canal due to the interaction with the suspended cohesive particles.

In the canal stretch between 0 and 8 km, sediment deposition increases with time, with most accumulation (1.5 m) in the upstream of the major canal, the deposition in the first 8 km of the major canal consists mostly of non-cohesive sediments (Figure 5-5). With sufficient flow velocity, the transport capacity is sufficient to convey the sediments along the major canal. Just after 8 km in the vicinity of the first two outlets (b1 and b2) sediment locally accumulate. In the Delft3D model, we specify the amount of water drawn by the outlets (0.5 m^3/s); the amount of suspended sediment removed by the outlet cannot be specified—it equals the amount of water withdrawn times the locally computed sediment concentration. After the outlet with less water remaining in the major canal, velocity and hence sediment transport capacity reduces leading to sediment deposition.

The sediment deposition gradually decreases until 9.1 km where the first weir and two outlets (b3 and b4) are located. One would expect the deposition to increase again due to the low velocity. However, due to the canal contraction close to these outlets the flow velocity increases. These two opposite effects more or less even each other out and the velocity remains approximately equal. As a result, the sediments continue to be moving downstream to 14.2 km. Just after 14.2 km, there is a big canal contraction causing erosion in the canal section upstream of the second weir due to the acceleration of flow velocity. The Toman minor canal and the last two outlets (b5 and b6) are located upstream of the second weir at 14.2 km. The outlets should decrease the velocity since they draw water from the major canal but because of the big canal contraction, velocity increases and erosion occurs. Thereafter the sediments continue to be transported till the end of the major canal.

In the minor canal, the gate is fully opened so the minor canal gets water carrying mostly cohesive sediments which deposit in the upstream of the minor canal (deposition reaches to 0.7 m). Since there is no structure disturbing their movement, the sediments are transported along the minor canal till the end (Figure 5-5), where the profile of the bed level shown is along the centerline (which is typically the deepest point of the cross-section). For more details, see the PowerPoint contains movies showing the updating of morphology within the cross-section at different locations in the major canal. The link for the supplementary data is: https://drive.google.com/drive/folders/1Wlw9SQSqGgRBLxyIoQ5FOjqdYAmxXFVV?usp=s haring.

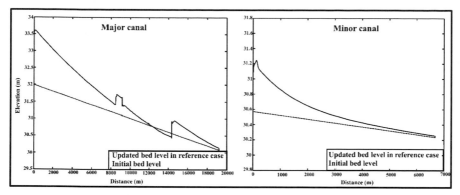

Figure 5-5 Sedimentation and erosion of sediments in the reference case.

The flow velocity gradually reduces along the minor and major canals except above the weirs explaining the sedimentation and erosion patterns along the canals and within the cross-section (Figure 5-6).

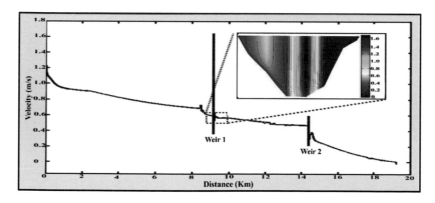

Figure 5-6 Flow velocity along the Zananda major canal and in the cross-section near the first weir.

The flow velocity (averaged over the cross-section) gradually reduces along the major canal (Figure 5-6). Within the cross-sections along the canal, the velocity distributions differ. For example, at the first weir, the average velocity is 0.6 m/s. The maximum velocity of 1.6 m/s occurs in the middle of the cross-section, while the velocity is at the sides is much less with 0 m/s close to both sidewalls. The left side has a higher velocity than the right side due to the asymmetric shape of the canal contraction and offtakes nearby (for more details, see the PowerPoint contains other movies showing the behaviour of velocity in the system near the diversion to minor canal. The link for the supplementary data is:

https://drive.google.com/drive/folders/1Wlw9SQSqGgRBLxyIoQ5FOjqdYAmxXFVV?usp=sharing).

Also along the minor canal, the average flow velocity drops from 0.5 m/s upstream of the gate to 0.21 m/s at the downstream (Figure 5-7). Likewise, the flow velocity distributions within the cross-section vary with the highest velocities in the middle and lower velocities on both sides due to the roughness of the wall. In the downstream of the canal, the velocity distribution is logarithmic where higher velocities at the top layer of water and lower velocity near the bed. In the upstream near the gate, the water flows underneath the gate and the top layer velocity became less than the bottom layer velocity (Figure 5-7).

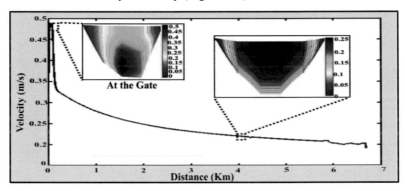

Figure 5-7 Velocity distribution in the Toman minor canal at different cross-sections.

Due to differences in velocity distribution, sediment is distributed in an asymmetric way within the cross-sections of the major and minor canal. The sediment behaviour is influenced by multiple factors such as the velocity, widening, and contractions of the canals and bed shear stress.

Figure 5-8 displays the difference in the deposition pattern between cohesive and non-cohesive sediments along the Zananda major canal. While cohesive sediments are gradually depositing along the major canal, the non-cohesive sediments are rapidly depositing in the first kilometers upstream of the canal with pronounced peaks and troughs in concentration near the weirs at 9 km and 14 km and canal contraction at 12.5 km. Non-cohesive sediments are deposited in the middle of the cross-section more than at both sides while the cohesive sediments are depositing almost equally in the middle and on both sides. In the case of pure cohesive sediments entering the irrigation system, most suspended sediments would be carried with the flow till the end of the major and minor canal. However, because in this case, the sediment is a mix of cohesive and non-cohesive, due to interaction with the heavier non-cohesive particles, some of the suspended cohesive particles start depositing in the upstream and middle of the canal stretches.

Figure 5-9 displays the difference in the deposition pattern between cohesive and non-cohesive sediments along the Toman minor canal. The same behavior will be there, where the cohesive sediments are gradually depositing along the minor canal, while the non-cohesive sediments are rapidly depositing at the beginning of the minor canal near the gate.

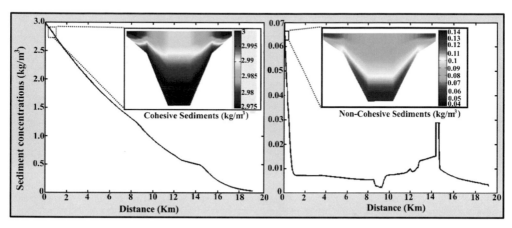

Figure 5-8 Distribution patterns of cohesive and non-cohesive sediments in Zananda major canal at different cross-sections.

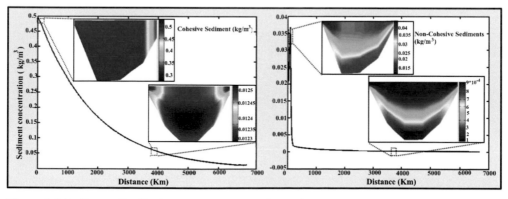

Figure 5-9 Sediment distribution patterns of cohesive and non-cohesive sediments in the Toman minor canal at different cross-sections.

The cohesive sediment concentration in the minor canal is much higher than the non-cohesive sediment concentrations. This is the opposite of the situation in the major canal.

The deposition pattern between cohesive and non-cohesive sediments is different in the minor canal. Where cohesive sediments are gradually depositing along the minor canal, vice versa for the non-cohesive sediments. The pattern of depositing of cohesive and non-cohesive sediments within the cross-section is the same as in the major canal, with the highest concentrations at the bottom and sides.

At the diversion to the minor canal, the velocity at the left side of the canal is reduced due to less water. Because of the subsequent reduction in velocity, a considerable amount of both non-cohesive and cohesive sediment is deposited, especially upstream the gate in the minor canal (Figure 5-10).

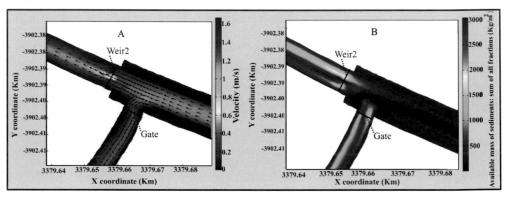

Figure 5-10 The relation between the velocity (A) and the amount of sediments (B) deposited at the diversion to the minor canal.

Figure 5-10 illustrates the effect of the velocity on the deposition and the transportation pattern of sediments. Panel (A) shows the reduced velocities at the right side of the canal after the diversion and the contraction. Panel (B) shows a higher deposition in these locations.

Acknowledging the asymmetric deposition patterns in the figures above, it can be noted the importance of using 2D/3D models to simulate sediment transport in the irrigation systems. Using Delft3D in this study proved useful in showing where the sediment is eroded or deposited and distributed along and within the cross-sections. Furthermore, Delft3D can show which kind of sediment is deposited where and in which quantities (Figure 5-11).

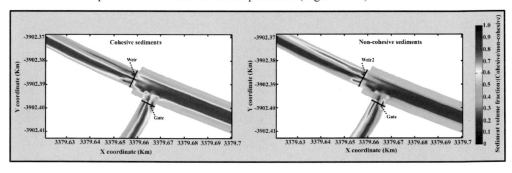

Figure 5-11 The difference in sediment distribution between the cohesive and non-cohesive sediments at the diversion.

The deposition and distribution of both kinds of sediments are different where the large amounts of cohesive sediments pass through the minor canal (Figure 5-11). On the other hand, less non-cohesive sediments enter the minor canal (Figure 5-11) since it is rapidly deposited in the upstream part (for more details, see the PowerPoint which contains movies showing the difference in distributions between the cohesive and non-cohesive sediment in the system, also movies showing the difference in the distribution of both sediments within the cross-section in the minor canal. The link for the supplementary data is: https://drive.google.com/drive/folders/1Wlw9SQSqGgRBLxyIoQ5FOjqdYAmxXFVV?usp=s haring).

5.3.2 Effect of weirs 1 and 2

Effect of the upstream weir (weir 1)

To see the effect of weir 1, we compare the sedimentation while reducing or raising the crest height of the weir. Raising the weir to 0.6 m increases the deposition slightly because of the obstruction of the water flow and creation of a backwater curve which leads to an increase in the water level and water depth. Combined with a constant discharge this leads to reduced velocity, reduced sediment transport capacity and hence more sediment deposition. This effect is noticeable only upstream of the weir and in the upstream part of the major canal, with negligible effect in the downstream part of the major canal (Figure 5-12).

Lowering or removing weir 1 leads to reduce deposition because water moves freely without structures disturbing its movement so the sediment transport capacity is sufficient to move sediments. The effect is noticeable upstream of the weir and in the upstream part of the major canal, with negligible effect in the downstream part of the major canal. The effect of the lowering or increasing the weir height has little effect on the minor canal (Figure 5-12).

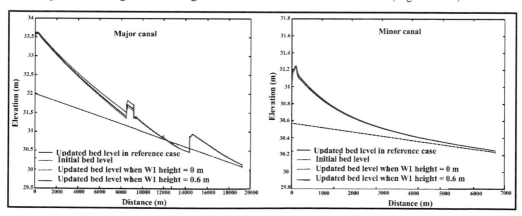

Figure 5-12 The effect of the upstream weir on sedimentation in the major canal (left panel) and the minor canal (right panel).

105

Changing the upstream weir settings reduces the sediment deposition significantly in the major canal while it has a negligible impact on the minor canal bed morphology (Figure 5-12).

Effect of the downstream weir (weir 2)

To evaluate the effect of weir 2, the weir has been raised and lowered in a similar way as weir 1, and compared the results with the reference case, the results shown in Figure 5-13 were too close. For this reason, changing the crest height of weir 2 has a little impact on sediment transport in the major and minor canals (Figure 5-13). Lowering and raising the downstream weir does not reduce the negative impacts of sedimentation, where the reduction in the deposition in both canals is very small.

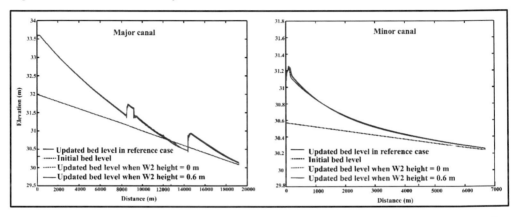

Figure 5-13 The effect of the downstream weir on sedimentation in the major canal (left panel) and the minor canal (right panel).

3.2.3. Effect of both weirs

In this scenario, we lower and raise both weirs simultaneously to see if there is a bigger impact on sediment transport. By comparing the results with results of the reference case, similar results were got as shown in Figure 5-12 and Figure 5-13.

5.3.3 Effect of gate settings

Constant gate opening

To see the effect of changing gate settings on sedimentation in the major and minor canal, the model was run with different gate openings to 0.2 m; 0.4 m; 0.6 m and 0.8 m and compared the modelling results with the reference case. Lowering the gate has a small impact on the major canal but substantially reduces sediment deposition in the minor canal. In case of gate openings equal to 0.2 m and 0.4 m sediment deposition almost fully blocks the canal reducing the flow into the minor canal to close to zero. The deposition in the first kilometers of the minor canal occurs due to the effect of weir 2. Due to the disturbance in flow caused by weir 2 the water entering the minor canal is well mixed and loaded with sediment. The backwater curve due to weir 2 causes sediment deposition (Figure 5-14). Lowering the gate reduces the deposition in the minor canal but at the same time, only a small amount of water can pass through the half-blocked canal which will not be sufficient to meet crops water requirements. Raising the gate can flush the sediment away.

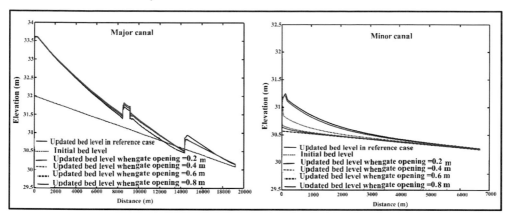

Figure 5-14 The effect of different fixed gate openings in time on sedimentation in the major canal (left panel) and the minor canal (right panel).

Figure 5-14 presents the effect of different fixed gate openings on the sediment deposition patterns in the major and minor canals. Reducing the gate height has a negligible impact on the major canal but a significant impact on sedimentation in the minor canal. However, reducing the gate also means less water entering the minor canal which may lead to insufficient water delivery to crops. Even though the gate setting of 0.8m reduces the sediments deposition less than the other gate settings compared to the reference case, the larger opening ensures sufficient water to meet crop water requirements. The large sediment deposition is located at the upstream part of the minor canal and the subsequent narrowing of the canal is visible in the field and on Google Earth imagery (Figure 5-15) providing further evidence that modelling results mimic the actual situation.

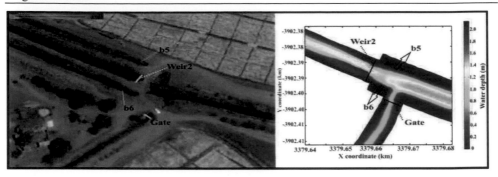

Figure 5-15 Comparing the Delft3D model results with actual field conditions as captured by Google Earth.

Variable gate openings following different operation plans

To test the impact of changing gate operation on the sedimentation in the canals, we formulate two different operation plans with different openings and time intervals based on the crop water requirements which change with crop growth stage. we prepare the first operation plan as shown in Figure 5-16 based on the data from (Osman, 2015). However, we simplified it by reducing the number of closing and opening the gate while keeping the same water distribution.

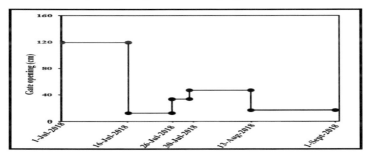

Figure 5-16 Operation plan with varying openings and time intervals based on crop water requirements.

The second operation plan is prepared, by fully closing and opening the gate at varying time intervals taking into account crop water requirements (Figure 5-17).

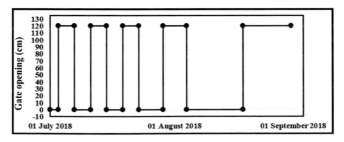

Figure 5-17 Operation plan with fully opening or closing the gate but with varying time intervals taking into account changing crop water requirements.

The first operation has a limited impact on sedimentation in the major and the minor canals as compared to the reference case. On other hand compared to the reference case operation plan 2 leads to a reduction in sediment deposition by half in the minor canal but limited impact on the major canal (Figure 5-18) while still meeting crop water requirements in a satisfactory manner. During the closure of the gate in the second operation plan the sediment accumulates near the gate and entrance to the minor canal. This is flushed away after fully opening the gate.

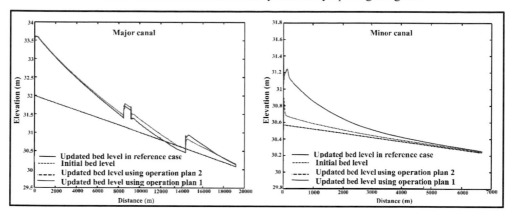

Figure 5-18 Effect of gate operation plan (in which the gate is either fully closed or opened at different time intervals) on sedimentation of the major canal (left panel) and left canal (right panel).

5.4 DISCUSSION

Many factors affect the flow and the sediment movement in the irrigation canals. The offtakes diverting to the branch canals and field outlets catering for different water requirement, the changes in the canal geometry (contraction or widening) and other parameters all affect hydrodynamic and morphologic parameters which determine canal performance and capacity to transport sediment. In this chapter, we illustrated how the location and the settings of weirs and gates do affect hydrodynamic and morphologic parameters.

Comparing scenarios to reduce sediment deposition in major and minor canals

Table 5-4 summarizes the results of the scenarios related to the impacts of the weir and gate settings on the amount of cohesive and non-cohesive sediment in the major and minor canal. The last column provides a qualitative assessment of whether the weirs and gate settings in the scenario can meet the crop water requirements (CWR), based on the quantity of water that can be delivered to the outlet.

Table 5-4 The impacts of operation in the sediment deposition as compared to the reference scenario

Scenario	Description		Major canal		Minor canal		Meeting
			Cohesive	Non-cohesive	Cohesive	Non-cohesive	CWR
Scenario 2 (Weir effects)	Weir 1	w1=0	−0.5%	−0.5%	No change	No change	3
		w1=0.6	1%	1%	2%	2%	2
	Weir 2	w2=0	No change	No change	3%	3%	2
		w2=0.6	No change	No change	−1.2%	−1.2%	3
	Both Weirs	w1=w2=0	−2%	−2%	3%	4%	2
		w1=w2=0.6	1%	1%	−1.1%	−1.1%	2
Scenario 3 (Gate effects)	Fixed gate height	g=0.2	4%	4%	Block	Block	0
		g=0.4	4%	4%	Block	Block	0
		g=0.6	3%	3%	PB*	PB*	1
		g=0.8	1%	1%	−19%	−19%	2
	Operation	Plan1	−0.1%	−0.1%	No change	No change	3
		Plan2	3%	3%	−54%	−55%	2

where (-) denotes a reduction and (+) an increase in sediments deposition as compared to the reference scenario; CRW = crop water requirement is assessed qualitatively in which 0 = no water, 1 = insufficient water for crops, 2= more or less sufficient water to satisfy CWR, 3 more than sufficient water to satisfy CWR, and PB=partially blocked.

The upstream weir (w1 in Figure 5-1) has some impact on the deposition in the upstream of the major canal while in the downstream part and in the minor canal the effect is negligible. Raising the weir height disturbs the water flow and creates a backwater curve which leads to an increased water level and with constant discharge reduced velocity, ultimately resulting in reduced velocity, reduced sediment transport capacity and deposition of sediments. The downstream weir (w2) has less impact on both canals. Simultaneously lowering or raising both weirs resemble the results of the individual weir settings.

The results of scenario 3 with fixed gate settings reveal a relatively small impact on sedimentation in the major canal but a potentially large impact on the minor canal. Lowering the gate less than 0.6 m leads to substantial sediment deposition at the entrance and upstream of the minor canal. The sediment deposition in the first kilometer of the minor canal occurs because 1) due to the flow disturbance caused by downstream weir, water entering the minor canal carries the eroded sediment and 2) due to the small gate opening, less water flows in the minor canal leading to lower sediment carrying capacity and hence deposition further downstream. Over time this leads to a complete or partial blockage of the minor canal which will adversely affect the capacity to deliver sufficient water to meet crop water requirements. It should be noted however that the Delft3D model may not be able to accurately model the local sedimentation near the gate and subsequent canal blockage. We used the 3D hydrostatic mode with a limited resolution which cannot resolve the full 3D details of the local flow near structures. Modelling the flow and sediment dynamics at an even higher resolution would be desirable but is out of scope for operational reasons (mostly due to significantly increased simulation times).

Compared to the reference case the two operation plans (both based on crop water requirements but one with variable gate settings, the other with variable time intervals) have a limited impact on the major canal. However, the second operation plan reduces the sediment deposition in the minor canal by more than 50%. In this operation plan during the closure of the gate, sediment is deposited upstream the gate; the subsequent full opening of the gate flushes the sediment away. This could be incorporated as a convenient maintenance practice.

Table 5-4 shows that the best operation scenarios are 1) fixed gate opening at 0.8 m where crop water requirement can be met in a satisfactory manner while reducing sediment deposition in the minor canal 2) operation plan 2 with either fully closing or opening the gate at variable intervals. Sediment accumulated during gate closure can be flushed away by fully opening the gate.

Osman et al. (2017) found in one of her operation scenarios that reduced inflows during the high sedimentation period by 51% led to sediment reduction of 48%. In this chapter in our first operation scenario, the same timings and gate settings was used as used by Osman but kept the flow and (high) sediment concentration constant. The effect on sedimentation in the canals is small. Hence we conclude that in Osman's scenario, the reduction of sediment-laden flow was the dominating factor in reducing sediment accumulation in the canal. Our second operation scenario shows the beneficial role of the intermittently opening and closing gate, even with high flow and sediment load. In this scenario, we assume constant sedimentation load temporally and spatially. In practice, sediment concentration in canals varies: some canals have very little or no sediments while others are suffering from high concentrations. Further, in some months sediment loads in the river are more severe than in others. Adjusting the timing of gate operation by closing the gate during periods of high sediment loads in the river to avoid sediments entering the minor canals can further reduce sediment problems.

The use of Delft3D for simulating sediment deposition in irrigation systems has significant advantages: 1) the 2D and 3D mode show where in the canal, longitudinal and in cross sections, deposition and erosion takes place; 2) the RTC feature allows for including weirs and gates that can be adjusted during the simulation, to mimic gate operations; 3) the model handles non-steady flow well. This is important in irrigation systems where structures (gates and weirs) in the canal disturb the water flow; 4) the model can handle both cohesive and non-cohesive sediment and their interaction. The latter is important where irrigation systems use natural rivers which typically carry a mix of sediments. The Delft3D model helps to understand the mechanism of sediment transport, to predict the location, quantity, and type of sediment accumulation under different operational scenarios. This information is essential for the design of operation and maintenance plans that will be effective in reducing sediment problems in irrigation systems.

As any other numerical model, Delft3D has limitations: 1) being developed for rivers, Delft3D does not simulate well the effects of sidewall roughness which makes the model inappropriate for narrow rectangular canals; 2) Delft3D and other 2D/3D hydrostatic models cannot predict local scour because vertical accelerations of the flow are ignored, turbulence modelling is limited, and the sediment transport formulations are based on smooth flow conditions. For local scour detailed 3D non-hydrostatic models are needed with non-equilibrium sediment transport pickup and deposition processes (Thanh et al., 2014); 3) Delft3D does not take into consideration the effect of consolidation of (cohesive) sediments and makes no distinction between newly deposited fluffy material and old consolidated materials (Zhou et al., 2016).

Finally there are two model implementation issues that need attention: 1) due to the much higher resolution than typical 1D models, simulation time can be extremely long, especially for large irrigation networks, despite useful tools such as Domain Decomposition, Flexible Mesh and Morphologic Factor, for example, in this study the cup time was 3.5 days and 2days for 3D and 2D modelling respectively; 2) the model implements the Q-h relationship as boundary condition in the downstream (i.e. water level as a function of the outflowing discharge Q). When the canal becomes dry and the water depth H drops to zero, this boundary does not reopen when the canal starts carrying water again. This situation frequently occurs in irrigation canals that are intermittently wet and dry depending on the water allocation plan. Most of these limitations are not insurmountable to solve since the model Delft3D is continuously developed further.

5.5 CONCLUSIONS

Efficient and well-executed canal operation plans can substantially improve hydraulic performance and reduce sediment problems which may lead to lower maintenance costs and as the result may increase crop production. This requires the proper operation of gates and finding the right balance between providing sufficient water for crop production and reducing sedimentation by the reduced sediment-laden flow. Our scenarios in the Gezira scheme in Sudan show how adjusting gate settings and varying timing of opening can be effective in reducing sedimentation in the secondary, distributary and field canals while meeting crop requirements in a satisfactory manner.

The Delft3D model, originally designed for rivers, was validated using measured field data from a previous study. The model was able to represent the actual condition (as shown in Figures 4 to 15). The biggest advantages of the model (as compared to previous sediment studies) proved its ability to model both cohesive and non-cohesive sediments and its 2D mode.

The latter allowed viewing flow parameters and sediments pattern within the cross-section, near offtakes, near gates and weirs and in the longitudinal direction. Determining the exact position of the sediment accumulation will help to reduce the maintenance costs and efforts and will also help the stakeholders to decide on the best operation to meet the crop water requirements while simultaneously minimizing sediment problems. Using a 3D model for cohesive and non-cohesive sediment, this study provides a substantial step forward in modelling the effect of structures on sediment behavior in irrigation canals and the use of gate operation to reduce sediment problems. Further studies are needed, in particular on the use of 3D models for large canal networks and with a better resolution around control and regulation structures. Running time and model stability are challenges here. Also, studies about the effect of gate operation with variable sediment concentrations will refine our scenarios.

Supplementary Materials: The PowerPoint and other helpful movies are available online at: https://drive.google.com/drive/folders/1Wlw9SQSqGgRBLxyIoQ5FOjqdYAmxXFVV?usp=sharing

6

CONCLUSIONS AND RECOMMENDATIONS

6.1 CONCLUSIONS

Sedimentation in irrigation canals can result in low irrigation performance by blocking canals, disrupting water supply, causing malfunction of structures and leading to high maintenance costs. Consequently, a number of studies have been conducted to simulate sedimentation and erosion in irrigation canals. However, those studies are mostly using 1D models and primarily deal with non-cohesive sediment. In reality sediments in irrigation canals consist of a mix of cohesive and non-cohesive sediments. To better understand the sedimentation patterns in canal networks 2D or 3D representation are needed.

This research used the Delft3D model to simulate the effect of canal operation on the cohesive and non-cohesive sediment transport and their deposition and erosion patterns in irrigation schemes. The Delft3D model was originally designed for rivers, estuaries and coastal areas. This model was chosen for this study because it is well documented and tested, it can handle cohesive and non-cohesive sediments and it is adaptable for use in canal systems. In this chapter, the most substantial findings of the study are presented.

The Delft3D model was applied in four steps of increasing complexity. Firstly, the model was tested on a simplified canal system, consisting of a 500 meter long main canal, one branch canal and several structures: one weir and a fully opened gate (Chapter 2). The model was tested for canals of different sizes, different shapes (rectangular and trapezoidal) and b/h ratio. After some adaptations, the Delft3D model was able to realistically represent water and sediment flows, from a hydrodynamic and morphologic point of view. The hydrodynamics were validated using the results from DUFLOW.

As a next step, the Delft3D was applied to simulate the effect of different types of sediments (cohesive and non-cohesive) in a range of different concentrations and mixtures, applying different flow discharges, in the same simplified canal set-up (Chapter 3). This provided insights in the differences in behaviour and deposition patters of cohesive and ono-cohesive sediments and their sensitivity to changes in concentration, and flow parameters. It also illustrated the effect of particle interaction on deposition patterns.

Thirdly, after establishing the use of Delft3D in a simplified canal network, the model was applied to an existing irrigation system in Nepal, the Sunsari Morang system (Chapter 4). In this application the impact of gate selection and operation on the deposition of non-cohesive sediments was investigated. The model was validated by data from previous studies and the sedimentation patterns observed in the settling basin.

Fourthly, one of the major and minor canal of the Gezira irrigation system in Sudan was used as case study to understand the effects of structures (offtake, weirs and gates) and different gate operation plans on the sedimentation patterns of mixed cohesive and non-cohesive sediments (Chapter 5). Several scenarios of gate operation options and weir heights were tested.

Other factors which are not addressed in this research like the discharge ratio, offtake location, offtake angle, state of any control structure, and sediment load characteristics all could have considerable influence the sediment distribution and resulting deposition patterns in the network. In order to study and quantify all these aspects would require a significantly larger number of simulations and therefore we have limited our conclusions to the basic impact trend of bigger effects on the more upstream main canals and smaller impact of the branch canals further downstream.

6.1.1 How can Delft3D be used in irrigation setting?

Delft3D has been used extensively in sediment simulation studies of rivers, estuaries and coastal zones. This study was the first documented attempt to apply it in irrigation systems. The model was applied to a simplified canal network to establish the feasibility for its use in canal networks, running it in 2D and 3D mode under different scenarios related to concentration of cohesive and non-cohesive sediments and canal discharges. Thereafter it was applied to two existing irrigation schemes in Nepal and Sudan. The Delft3D model has several advantages over other 3D models used in the simulation of sediment flows, such as Mike and Telemac. The most important reason to choose Delft3D for this study was its possibility to simulate both cohesive and non-cohesive sediment and a mixture of sediments. Furthermore, Delft3D comes with several in-built tools such as Flexible Mesh, Domain Decomposition and Real-time Control, which proved useful for its use in canal networks and for simulating structures and gate operation.

The applications done in this study clearly the feasibility and usefulness of using Delft3D in irrigation settings. The model applications gave satisfactorily results, except for the smaller canals because of the sensitivity of Delft3D towards side wall friction.

6.1.1.1 *Adaptations of Delft3D for use in irrigation canals*

After applying and calibrating Delft3D three limitations were found that are specific to the application in irrigation settings (as opposed to rivers and estuaries). These are:

- In the simulations the Q-h relationship is used as boumdary condition. This is reversed in Delft3D as compared to irrigation models which typically use h-Q relationship. Consequently the Delft3D model can not deal with zero water depth. In rivers this is a lesser problem, but in irrigation canals water depths are frequently zero when they are not carrying water.

- Side wall friction: the results show that Delft3D is very sensitive to the roughness of the canal sides, especially for narrow rectangular canals with b/h ration less than 3.

- The long simulation time: In the Delft3D model structured grids with rectangular cells and areas are computationally efficient if aligned with long straight canals. However, an irrigation system consists of main canals and branches, with large 'empty' areas in between. These 'empty' inactive parts, which fall outside the area of interest, render the structured grids inefficient since the model domain includes large inactive parts taking up unnecessary computation time, and leading to long simulation time.

To address these issues the following adaptations were made to the model:

- The Q-h relationship was accommodated by creating a virtual drop structure of 0.5 m at the downstream end of each of the canals, in order to avoid the dry flow (zero water depth) in the model that causes the model to stop running, and in order to dissipate the energy caused by the sudden change in elevation without producing a scour in the canal, avoiding the critical flow by lowering the water elevation while allowing a subcritical flow in the main canal.

- For the second limitation, the sidewall friction, I adapted the model by lowering the sidewall roughness to 10 times less than the bed roughness and acceptable results were obtained.

- The third limitation, the long simulation time, was addressed using a tool in Delft3D called the "Domain decomposition" which reduced the simulation time by 60%. Secondly, by using the refinement property where I chose big grid cells for the long straight canal and small grid cells for areas of interest like bends, structures, and offtakes the simulation time was reduced to 40%. Thirdly, choosing a higher performance efficient computer played a significant role in reducing the computation periods.

6.1.1.2 *The outputs of Delft3D through the research*

After these adaptations in the model, the primary outputs of using Delft3D in irrigation canals were encouraging. The Delft3D model was able to provide a realistic image about the velocity distribution along the canals as well as within the canal cross-sections. From a hydrodynamic point of view, it is concluded that Delft3D can be used in irrigation systems simulations since it can provide good and realistic results. From the morphodynamic point of view, also it is concluded that Delft3D can be used in irrigation systems that suffer from sediments problems since it provided realistic results for cohesive sediment behaviour in the horizontal and vertical direction.

Then the Delft3D model was used to show the differences between cohesive and non-cohesive sediments. Delft3D model was very useful in the simulations to distinguish the differences between the cohesive sediments, non-cohesive sediments and their mixture. The deposition and erosion patterns of the sediments are not uniform along the canal and they differ from one location to another. Also in the cross-sections sediments are distributed in a non-evenly way especially near gates, weirs, offtakes, and diversions. Delft3D was able to show the differences between the two kinds of sediments regarding their behaviour, the locations of their distribution, their deposition and erosion patterns, their sensitivity towards different flow and sediments parameters, and the velocity distribution along the canals and within their cross-sections, other flow parameters as well. Lastly, Delft3D shows evidently the interaction between the two kinds of sediment and how they affect each other.

Then Delft3D was applied to a case study in Sunsari Morang scheme in Nepal. Delft3D was used to simulate the non-cohesive sediments to study the effect of the selection of the gates on the non-cohesive sedimentation. The use of the Delft3D model was very helpful in understanding and visualizing the hydrodynamic as well as morphologic parameters. It shows clearly how the selection of different gates affects the flow distribution, the amount of water diverted to the branch canal, the amount of sediment to be transported to the branch canal and to be deposited in the main canal, how the bed of the canal is formed due to different velocities which may cause deposition or erosion, also the sediments distribution along the canal and within the cross-sections especially near the gates. Using a 1D-model would not detect any difference in the gate a, b, c or d, while 2D/3D models as Delft3D show exactly the location of the deposition which is different.

Lastly Delft3D was used in another real irrigation system to simulate the mixed sediments " mostly cohesive sediments'' in Gezira scheme in Sudan to study the effect of different structures on the cohesive sedimentation and movement and to study the effect of gate operation on these sediments' distribution. Delft3D-4 suite proved a useful tool in this analysis, where Delft3D helped in providing better insights into the sediment transport patterns and spatial distribution of deposition and erosion along the canal and within cross-sections. This study provides a substantial step forward in the sediment transport modelling in irrigation systems with 2D and/or 3D perspective.

6.1.1.3   *The use of 1D versus 2D & 3D*

In order to study hydrodynamics in irrigation canals, using 1D models is computationally efficient However, these models are not representative in morphologic simulations. 1D-models have a limited capacity to present many of the important three-dimensional phenomena found in nature. These models can only present the quantities in longitudinal direction.

On the other hand, 2D/3D models are more efficient from a morphological point of view, since they can detect sediment transport and sediment patterns near offtakes and structures in more details, and simulate deposition and/or erosion locations within the cross-section in addition to those in the longitudinal direction.

In this research, the 2D/3D was able to provide a realistic image of velocity distribution along the system and in canal cross-sections. The use of the 2D/3D model was desirable for the simulation of sediments since the deposition patterns are not uniform along the canals and are not evenly distributed in the canal cross-sections. In particular, near offtakes, diversions, and canal structures, where the use of the 2D model was very helpful in understanding and visualizing the hydrodynamic as well as morphologic parameters and their distribution along the canal and within the cross-section. Determining the exact position of the sediment accumulation will help to reduce the maintenance costs and efforts and will also help the stakeholders to decide on the best operation to meet the crop water requirements while simultaneously minimizing sediment problems.

Using different gates give a similar quantity of sediments, but the location in the canal is very different. Using a 1D-model would not detect any difference in the use of which gate. On the other hand 2D-models show the location of the deposition/erosion within the cross-sections. 3D-models may be more beneficial and may give better results but the running time is too long and computationally challenging.

However, there is a limitation that needs attention which is the long simulation time, due to the much higher resolution than typical 1D models, simulation time can be enormously long, especially for large irrigation networks, despite useful tools such as Domain Decomposition, Flexible Mesh and Morphologic Factor.

Despite using 2D/3D simulations, I can represent the outputs of the model in 1D, 2D, and 3D figures, where 1D figures can represent different parameters in the longitudinal direction for long straight canals. While 2D figures can present the plan view of the system, especially when representing the dynamics near the diversion to the branch canals. The 3D figures can present the flow dynamics as velocity distribution, also the bed level development within the cross-sections. While challenges regarding the use of 2D/3D simulation models in canal systems remain, the results in this study show that it is a step forward from previous studies.

6.1.2 Cohesive and non-cohesive sediments

Most of the previous studies in irrigation schemes were done regarding non-cohesive sediments. My second research question was: How will cohesive sediments differ from the non-cohesive sediments and their mixture regarding their distribution, canal bed morphology development, their sensitivity, and deposition and erosion in different locations? How do cohesive and non-cohesive sediments interact?

Cohesive and non-cohesive sediments behave very differently due to particle size, shape, weight and ionic charge. Cohesive and non-cohesive sediments are different in their distribution such as locations of deposition and erosion, movement as suspended or rolling. They are different in their sensitivity towards the changes in flow and sediment parameters. The interaction between the two kinds of sediments in mixtures lead to differences in behaviour. The deposition rate of non-cohesive sediment is lower in the mixed sediments than in the pure non-cohesive sediments, which means that non-cohesive sediments deposit slower in a mixture with cohesive sediment than in case of pure non-cohesive sediment of the same concentration.

Those big differences between cohesive, non-cohesive sediments and mixed sediments, should be considered and should not be ignored by assuming that there are no differences and assuming that the cohesive or mixed sediments can be simulated by using non-cohesive sediments tools.

Most of the studies on cohesive and non-cohesive sediments are done in river systems. There are fundamental differences between rivers and canal systems which influence the behaviour of sediments, for example, b/h ratio, sidewall friction, weirs and gates. The impact of these differences for cohesive and non-cohesive sediments are described in the paragraphs below.

Comparing the sediment behaviour in rivers and in irrigation canals: similarities and differences

Similarities

- Just like rivers, there is an interaction between cohesive and non-cohesive sediments in irrigation systems, where the cohesive sediment reduces the deposition rate of non-cohesive sediment as compare to pure non-cohesive sediments.

- The pattern of mixed sediments more closely resembles that of the pure non-cohesive than the pure cohesive scenario, indicating that, although there is an interaction between the two types, the behaviour of non-cohesive sediments is dominant over that of cohesive sediments. As the cohesive fraction increases, the behaviour starts to resemble more cohesive.

- Just like rivers, the canal bed shear stress in irrigation systems increases when sediments enter the system. Shear stress in case of non-cohesive sediment entrance is higher than in the case of cohesive.

- Similar to rivers, cohesive sediment is carried and moves in suspension. On the other hand, the deposited non-cohesive sediments rolling in bed as bedload, while the finer particles of the non-cohesive sediments move in suspension.

- Because of the low deposition rate of cohesive sediment, it takes a long time to deposit so most of the cohesive sediments are carried further. While for the non-cohesive sediment, the deposition rate is relatively high, so they are rapidly deposited at the upstream of the main canal.

- Cohesive sediment is more sensitive to variations in discharge, velocity, and shear stress compared to non-cohesive sediments. While the non-cohesive sediment is more sensitive to variations in sediment concentrations compared to the cohesive sediments. The non-cohesive sediments are sensitive to canal slope, settling velocity and the particle size.

Differences

Sidewall friction

Rivers are not affected by sidewalls they have negligible impact on the river flow, only the roughness for the bed is calculated. While in irrigation canals, the sidewall roughness plays an important role in the hydrodynamics of the canals as velocity distribution which affects the sediment transport through the canal, where lower velocities are found at the sides of the canal due to the friction. Delft3D is sensitive to the sidewall roughness, especially for small rectangular canals.

In irrigation canals, most of the cohesive sediments are deposited at both sides which besides the weed growth hinder the flow causing a reduction in velocity, while most of the non-cohesive sediments are deposited in the middle.

In rivers, most of the cohesive sediments are deposited in the middle while most of the non-cohesive sediments are deposited at sides of rivers.

pH

Generally, the accepted value of pH for irrigation water is between 5.5 and 7.5, but some problems can occur within this range. While the acidity of freshwater is naturally variable, but most lakes and rivers have a pH within the range of 6 to 9. Although any pH less than 7 can technically be considered acid

The initial suspension before settling at a lower pH (pH $\leq$ 5.5) will be flocculated. The resulting settled beds show strong stratified sediment bed offering a high erosion rate near the surface (Ravisangar et al., 2005). An intermediate pH ($5 \leq$ pH ≤ 7) leads to having a weaker bed structure due to lack in surface contact, which leads to more susceptibility to erosion. At high pH values when pH > 7, the surface attraction forces become significant and form denser aggregates. (Ravisangar et al., 2005) observed that acidic water produces stronger sediment, which it does not erode easily due to excess shear stress.

Generally, in rivers the salinity influences the cohesive sediments due to their ionic charges and leads to creates flocs of the cohesive sediments, these flocs are difficult to be removed. Saline water is rarely present in irrigation canals, so the probability of having flocs of cohesive sediment is weak. On the other hand, non-cohesive sediments are not affected by salinity. However, in this research, I didn't include pH in our simulation due to its negligible presence in irrigation systems.

6.1.3 The impact of gates operation on the sediment transport in the irrigation schemes

To answer the third research question regarding the impact of gate operation, I applied Delft3D to the Sunsari Morang Irrigation Scheme in Nepal which is suffering from non-cohesive sediments, primarily in the upstream of the main canal. I also applied Delft3D to simulate sedimentation in the Gezira Scheme in Sudan which is suffering from mixed sediments (cohesive sediments mostly).

I conclude the following:

- The location of the gate to be operated has a major influence in the hydraulic as well as sediment erosion and deposition pattern, where it was noticed that, opening the gate in the side near to the offtake resulted in sediment deposition at the entrance of the offtake less than when opening the gate in the side far from the offtake.

- The opening heights have an impact in reducing the undesirable sediment entering the branch canals.

- Number of gates to be operated, it has been noticed that when opening only one gate diverts more sediment to the branch canal than opening two gate or more gates, and less sediment will be transported to the end of the main canal.

- Sequence of gate operation

    If the same gate is opened regularly, then the deposition and erosion due to the gate operation can permanently change the canal's cross-section geometry and that is not preferable. Also, it noticed that the sediment once deposited during one gate operation can be eroded during the other gate operation and the sediments can be flushed out.

- Location of weirs can influence their effectiveness

- Operation plan of closing and fully opening the gate has a major impact on the minor canals.

6.2 REFLECTION

From an irrigation point of view, the use of well-designed 2D / 3D models can ultimately help understand sediment behaviour in irrigation canals and uncover precise locations with sediment problems. In addition, it helps to know the differences between the two types of sediment and the factors that increase/reduce their sediment. Knowing these differences can help in choosing the appropriate method to solve the sediment problem. Different methods (gate operation, canal rehabilitation, design, and canal modernization) can all help to reduce sediment. These methods can be chosen based on the modelling outputs. By solving sediment problems in irrigation systems, canal performance may be improved and crop production increased. For this reason, the Delft3D model is a useful tool for stakeholders and gate operators to create a relative analysis of sediment transport, by addressing differences between cohesive and non-cohesive sediments. In addition, by addressing the effects of different operating methods of sedimentation, as well as through access to the most effective sediment and water management. Further development of Delft3D to overcome its limitations with respect to long simulation time.

6.3 RECOMMENDATION

This research investigates how canal operation can help to prevent deposition and flush out already deposited sediments. The results of this research point to some recommendations:

- The use of the 2D/3D models is recommended since the deposition patterns are not uniform along the canals and are not evenly distributed. Delft3D 4 has provided promising results, and was very helpful in our analysis. However, Delft3D FM Suite is recommended, which allow us to reduce the simulation periods ''which was one of the limitations'' by using the 1D model for long straight canals and 2D & 3D models in the area of interest.
- From an operation point of view reducing water supply to half of the designed discharge is not recommended for canals which are suffering from non-cohesive deposition because discharge has a big impact on canal morphology development while reducing the water supply in canals which are suffering from cohesive deposition has a negligible impact.
- Operation plans which include variable openings, well-trained operators and effective maintenance of irrigation schemes are needed to minimize sediment problem and improve the hydraulic performance of canals which leads to increase crop yields and ensure sustainable production. For variable sediment concentration, using the operation plan of fully closing and/or fully opening the gate is vital, since I can close the gate when high concentrations enter the canal and open when fewer concentrations there.
- There is a great need to undertake research about the cohesive sediments in irrigation systems especially in big networks and about the operation effect on the sediment accumulation in the canals and around the control and regulation structures.

6.4 RESEARCH CONTRIBUTIONS

- The first contributions of this research was introducing and applying the Delft3D model in the irrigation system simulation, where this model was very important in showing the non-uniform flow and sedimentation patterns around offtakes and structures in the canal, also this model presented clearly the major differences between the cohesive and non-cohesive sediments and their mixture, finally, this model showed the impacts of different structures on sediment transport and showed the gate operation and gate selection to be operated impacts on the sediment transport and patterns of deposition and/or erosion.

- The second contribution was finding the interaction between cohesive sediments and non-cohesive sediments in irrigation systems.

- The third contribution was showing the major differences between cohesive sediments, non-cohesive sediments and their mixture in irrigation systems, and showing their sensitivity towards different flow and sediments parameters.

- A fourth contribution, building on earlier work, was confirming the importance of operating gates with different operation plans to reduce the negative impacts of sediments while meeting the crop water requirements.

- The fifth contribution was showing the relevancy of selecting gates to be operated on the sediment accumulation/ erosion and how by changing gates will help in the flushing of the sediments in order to reduce the maintenance costs.

6.5 FURTHER STUDIES

In order to enhance the knowledge realted to sediment transport in irrigation systems, there are some for further studies :

- Variable cohesive/ non-cohesive sediments concentrations;

- Complex networks and structures;

- Effect of variable discharges.

7

REFERENCES

Belaud, G. and Baume, J. P. (2002). Maintaining equity in surface irrigation network affected by silt deposition. Journal of Irrigation and Drainage Engineering, 128(5), 316-325.

Bhutta, M. N., & Shahid, B. A., and Van Der Velde, E. J. (1996). Using a hydraulic model to prioritize secondary canal maintenance inputs: results from Punjab, Pakistan. Irrigation and Drainage Systems, 10(4), 377-392.

Caviedes-Voullième, D., & Morales-Hernández, M., & Juez, C., & Lacasta, A., and García-Navarro, P. (2017). Two-dimensional numerical simulation of bed-load transport of a finite-depth sediment layer: applications to channel flushing. Journal of Hydraulic Engineering, 143(9), 04017034.

Celik, I. and Rodi, W. (1988). Modeling suspended sediment transport in nonequilibrium situations. Journal of Hydraulic Engineering, 114(10), 1157-1191.

Chan, W., & Onyx, W., and Li, Y. (2006). Critical shear stress for deposition of cohesive sediments in Mai Po. Journal of Hydrodynamics, Ser. B, 18(3), 300-305.

Chow, T. V. (1959). Open-channel hydraulics (Vol. 1): McGraw-Hill New York.

Clemmens, A., & Bautista, E., & Wahlin, B., and Strand, R. (2005). Simulation of automatic canal control systems. Journal of Irrigation and Drainage Engineering, 131(4), 324-335.

Cole, P. and Miles, G. V. (1983). Two-dimensional model of mud transport. Journal of Hydraulic Engineering, 109(1), 1-12.

De Jong, J. (2005). Modelling the influence of vegetation on the morphodynamics of the river Allier: MSc thesis. Technology university of Delft,

Deltares. (2016). Delft3D-Flow user manual. Available online: https://oss.deltares.nl/documents/183920/185723/Delft3D-FLOW_User_Manual.pdf (accessed on 28-5-2016)

Department of Irrigation. (2003). Design Report Vol-I, Main Report and Appendices. Prepared by NEDECO for Sunsari Morang Irrigation Project Stage III (phase-I), Ministry of Water Resources, Nepal.

Depeweg, H. and Méndez, N. (2007). A New Approach to Sediment Transport in the Design and Operation of Irrigation Canals: UNESCO-IHE Lecture Note Series: CRC Press.

Depeweg, H. and Paudel, K. (2003). Sediment transport problems in Nepal evaluated by the SETRIC model. Irrigation and Drainage, 52(3), 247-260.

Depeweg, H., & Paudel, K. P., and Méndez, N. (2014). Sediment transport in irrigation canals: a new approach: CRC Press/ Balkema.

Devkota, L., & Crosato, A., and Giri, S. (2012). Effect of the barrage and embankments on flooding and channel avulsion case study Koshi River, Nepal. Rural Infrastructure 3 (3), 124-132.(2012).

DFID, M. M. (2006). Equity, Irrigation and Poverty, Guidelines for Sustainable Water Management, Final Report.

Dulovičová, R. and Velísková, Y. (2009). Aggradation of irrigation canal network in Žitný Ostrov, Southern Slovakia. Journal of Irrigation and Drainage Engineering, 136(6), 421-428.

Elias, E., & Walstra, D., & Roelvink, J., & Stive, M., and Klein, M. (2001). Hydrodynamic validation of Delft3D with field measurements at Egmond. Paper presented at the Coastal Engineering Conference.

Flokstra, C. (2006). Modelling of submerged vanes. Journal of Hydraulic Research, 44(5), 591-602.

Flokstra, C., & Jagers, H. R. A., & Wiersma, F. E., & Mosselman, E., and Jongeling, T. H. G. (2003). Numerical modelling of vanes and screens; development of vanes and screens in Delft3D-MOR. Retrieved from Delft University of Technology: http://resolver.tudelft.nl/uuid:304fa85d-bc41-46d3-948c-0e91bd91a6ab

Gebrehiwot, K. A., & Haile, A. M., & De Fraiture, C., & Chukalla, A. D., and Tesfa-alem, G. E. (2015). Optimizing flood and sediment management of spate irrigation in Aba'ala Plains. Water resources management, 29(3), 833-847.

Gismalla, Y. A. (2009). Sedimentation problems in the Blue Nile reservoirs and Gezira scheme: a review. Gezira Journal of Engineering and Applied Sciences, 4(2), 1-12.

Guan, W. B., & Wolanski, E., and Dong, L.-X. (1998). Cohesive sediment transport in the Jiaojiang River estuary, China. Estuarine, Coastal and Shelf Science, 46(6), 861-871.

Huang, & Jianchun, & Hilldale, R., and Greimann, B. (2006). Cohesive sediment transport. Erosion and sedimentation manual, 1-46.

Huang, J., & C., H. R., and P., G. B. (2008). Erosion And Sedimentation manual. Denver, Colorado U.S.: Department of the Interior Bureau of Reclamation.

Hung, M., & Hsieh, T., & Wu, C., and Yang, J. (2009). Two-dimensional nonequilibrium noncohesive and cohesive sediment transport model. Journal of Hydraulic Engineering, 135(5), 369-382.

Javernick, L., & Hicks, D., & Measures, R., & Caruso, B., and Brasington, J. (2016). Numerical modelling of braided rivers with structure-from-motion-derived terrain models. River Research and Applications, 32(5), 1071-1081.

Jian, H. U. (2008). Study on mathematical modeling for non-uniform sediment transport in an irrigation district along the lower reach of the Yellow River. Journal of China Institute of Water Resources and Hydropower Research, 1, 005.

Jinchi, H., & Zhaohui, W., and Qishun, Z. (1993). A study on sediment transport in an irrigation district. Paper presented at the 15th International Congress on Irrigation and Drainage, the Hague, the Netherlands 4-11 September.

Kemp, L. (2010). The evolution of sandbars along the Colorado River downstream of the Glen Canyon Dam: MSc thesis. Delft University of Technology, Delft.

Kondolf, G. M., & Gao, Y., & Annandale, G. W., & Morris, G. L., & Jiang, E., & Zhang, J., & Cao, Y., & Carling, P., & Fu, K., and Guo, Q. (2014). Sustainable sediment management in reservoirs and regulated rivers: Experiences from five continents. Earth's Future, 2(5), 256-280.

Krishnappan, B. G. (2000). Modelling cohesive sediment transport in rivers. IAHS Publication(International Association of Hydrological Sciences)(263), 269-276.

Krone, R. B. (1962). Flume studies of the transport of sediment in estuarial shoaling processes: University of California.

Lawrence, P., & Spark, and Counsell, C. (2001). Procedure for the selection and outline design of canal sediment control structures. HR Wallingford and Department for International Development (DFID).

Lawrence, P. A., Edmund. (1998). Deposition of fine sediments in irrigation canals. Irrigation and Drainage Systems, 12(4), 371-385. doi:10.1023/a:1006156926300

Lesser, G. R. (2009). An approach to medium-term coastal morphological modelling: UNESCO-IHE, PhD Thesis: CRC Press/ Balkema.

Lesser, G. R., & Roelvink, J. v., & Van Kester, J., and Stelling, G. (2004). Development and validation of a three-dimensional morphological model. Coastal engineering, 51(8-9), 883-915.

Li, L. (2010). A fundamental study of the Morphological Acceleration Factor. Civil Engineering and Geosciences. Retrieved from http://resolver.tudelft.nl/uuid:2780f537-402b-427a-9147-b8652279a83e, 23-8-2010

Liu, W. C., & Hsu, M. H., and Kuo, A. Y. (2002). Modelling of hydrodynamics and cohesive sediment transport in Tanshui River estuarine system, Taiwan. Marine Pollution Bulletin, 44(10), 1076-1088.

Lopes, J., & Dias, J., and Dekeyser, I. (2006). Numerical modelling of cohesive sediments transport in the Ria de Aveiro lagoon, Portugal. Journal of Hydrology, 319(1), 176-198.

Luijendijk, A. (2001). Validation, calibration and evaluation of Delft3D-FLOW model with ferry measurements.

Mendez, N. J. (1998). Sediment transport in irrigation canals: UNESCO-IHE, PhD Thesis: CRC Press/ Balkema.

Mishra, S. K. (2016). Draft Report On Main Irrigation Canal Operation Plan For Crops In Sitagunj Secondary Canal (S9) Irrigation System. IWRMP, DOI, Nepal.

Morianou, G. G., & Kourgialas, N. N., & Karatzas, G. P., and Nikolaidis, N. P. (2016). Hydraulic and sediment transport simulation of Koiliaris River using the MIKE 21C model. Procedia Engineering, 162, 463-470.

Munir, S. (2011). Role of Sediment Transport in Operation and Maintenance of Supply and Demand Based Irrigation Canals: Application to Machai Maira Branch Canals: UNESCO-IHE, PhD Thesis: CRC Press/ Balkema.

Nawazbhutta, M., & Shahid, B. A., and Van Der Velde, E. J. (1996). Using a hydraulic model to prioritize secondary canal maintenance inputs: results from Punjab, Pakistan. Irrigation and Drainage Systems, 10(4), 377-392.

Nicholson, J. and O'Connor, B. A. (1986). Cohesive sediment transport model. Journal of Hydraulic Engineering, 112(7), 621-640.

Nippon Koei. (1995). Project Operation Plan. Sunsari Morang Irrigation Project, Ministry of Water Resources, Nepal.

Osman, & S., I., & Schultz, B., & Osman, A., and Suryadi, F. (2017). Effects of different operation scenarios on sedimentation in irrigation canals of the Gezira Scheme, Sudan. Irrigation and Drainage, 66(1), 82-89.

Osman, I. (2015). Impact of improved operation and maintenance on cohesive sediment transport in Gezira Scheme, Sudan: UNESCO-IHE, PhD Thesis: CRC Press/ Balkema.

Osman, I. S., & Schultz, B., & Osman, A., and Suryadi, F. (2016). Simulation of Fine Sediment Transport in Irrigation Canals of the Gezira Scheme with the Numerical Model FSEDT. Journal of Irrigation and Drainage Engineering, 142(11).

Osman, I. S. E. (2012). Sediment and Water Management of Large Irrigation Systems, Case Study: Gezira Scheme, Sudan. (PhD proposal), UNESCO-IHE institute, Delft.

Parsapour-Moghaddam, P. and Rennie, C. D. (2017). Hydrostatic versus nonhydrostatic hydrodynamic modelling of secondary flow in a tortuously meandering river: Application of Delft3D. River research and applications, 33(9), 1400-1410.

Partheniades, E. (1965). Erosion and deposition of cohesive soils. Journal of the Hydraulics Division, 91(1), 105-139.

Partheniades, E. (1986). A fundamental framework for cohesive sediment dynamics. In Estuarine cohesive sediment dynamics (pp. 219-250): Springer.

Partheniades, E. (2009). Cohesive Sediments in Open Channels: Erosion, Transport and Deposition. United Kingdom: Butterworth-Heinemann.

Paudel, K. P. (2010). Role of Sediment in the Design and Management of Irrigation Canals: Sunsari Morang Irrigation Scheme, Nepal: UNESCO-IHE, PhD Thesis: CRC Press/Balkema.

Paudel, K. P., & Depeweg, H., and Méndez, N. (2014). Sediment transport in irrigation canals: a new approach: CRC Press.

Ravisangar, V., & Sturm, T., and Amirtharajah, A. (2005). Influence of sediment structure on erosional strength and density of kaolinite sediment beds. Journal of Hydraulic Engineering, 131(5), 356-365.

Renault, D., & Facon, T., and Wahaj, R. (2007). Modernizing Irrigation Management: The MASSCOTE Approach--Mapping System and Services for Canal Operation Techniques (Vol. 63): Food & Agriculture Org.

Renault, D. and Wahaj, R. (2006). MASSCOT: a methodology to modernize irrigation services and operation in canal systems. Applications to two systems in Nepal Terai: Sunsari Morang Irrigation System and Narayani Irrigation System. FAO. Rome, 1-42.

Roelvink, D., & Boutmy, A., and Stam, J.-M. (1998). A simple method to predict long-term morphological changes. Coastal Engineering Proceedings, 1(26).

Roelvink, J. and Van Banning, G. (1995). Design and development of DELFT3D and application to coastal morphodynamics. Oceanographic Literature Review, 11(42), 925.

Sanford, L. P. and Halka, J. P. (1993). Assessing the paradigm of mutually exclusive erosion and deposition of mud, with examples from upper Chesapeake Bay. Marine Geology, 114(1-2), 37-57.

Schaffner, J. (2008). Numerical investigations on the function of flush waves in a reservoir sewer. Technische Universität, Darmstadt, Germany.

Schultz, B. (2002). Land and Water Development. Delft, the Netherlands: IHE.

Schultz, B. and De Wrachien, D. (2002). Irrigation and drainage systems research and development in the 21st century. Irrigation and Drainage: The journal of the International Commission on Irrigation and Drainage, 51(4), 311-327.

Sherpa, K. (2005). Use of Sediment Transport Model SETRIC in an Irrigation Canal. Unesco-IHE, Delft, Netherlands.

Simons, D. and Fuat, S. (1992). Sediment Trasport Technology. Colorado, USA.

Sutama, N. (2010). Mathematical modelling of sediment transport and its improvement in Bekasi Irrigation System, WestJava, Indonesia. MSc thesis Unesco-IHE.

Teisson, C. M. O., P. Le Hir, C. Kranenburg, L. Hamm,. (1993). Cohesive sediment transport processes,. Coastal Engineering, (http://www.sciencedirect.com/science/article/pii/037838399390048D), Volume 21(Issues 1–3), 34 Pages. doi:https://doi.org/10.1016/0378-3839(93)90048-D.

Thanh, N. V., & Chung, D. H., and Nghien, T. D. (2014). Prediction of the local scour at the bridge square pier using a 3D numerical model. Open Journal of Applied Sciences, 4(02), 34.

Theol, & A, S., & Jagers, B., & Suryadi, F., and de Fraiture, C. (2019). The use of Delft3D for Irrigation Systems Simulations. Irrigation and Drainage, 68(2), 318-331. doi:10.1002/ird.2311

Theol, S., & Bert Jagers, & F. Suryadi, and Charlotte De Fraiture. (2019a). The role of operation in reducing problems with cohesive and non-cohesive sediments in irrigation canals. MDPI/Water Journal.

Theol, S., & Bert Jagers, & F. Suryadi, and Charlotte De Fraiture. (2019b). The use of Delft3D for Irrigation Systems Simulations. Irrigation and Drainage, 68(2), 318-331.

Theol, S., & Bert Jagers, & F. Suryadi, and Charlotte De Fraiture. (forthcoming). The use of 2D/3D models to show the differences between cohesive and non-cohesive sediments in irrigation systems. Submitted to the American Journal of Irrigation and Drainage Engineering (ASCE).

Van der Wegen, M., & Jaffe, B., and Roelvink, J. (2011). Process-based, morphodynamic hindcast of decadal deposition patterns in San Pablo Bay, California, 1856–1887. Journal of Geophysical Research: Earth Surface, 116(F2).

Van Leussen, W. (1994). Estuarine macroflocs and their role in fine-grained sediment transport: Ministry of Transport, Public Works and Water Management, National Institute for Coastal and Marine Management (RIKZ).

Van Rijn, L. C. (1993). Principles of sediment transport in rivers, estuaries and coastal seas (Vol. 1006): Aqua publications Amsterdam.

Van Rijn, L. C., & Van Rossum, H., and Termes, P. (1990). Field verification of 2-D and 3-D suspended-sediment models. Journal of Hydraulic Engineering, 116(10), 1270-1288.

Villaret, C., & Hervouet, J.-M., & Kopmann, R., & Merkel, U., and Davies, A. G. (2013). Morphodynamic modeling using the Telemac finite-element system. Computers & Geosciences, 53, 105-113.

Winterwerp, J. C. and Van Kesteren, W. G. (2004). Introduction to the physics of cohesive sediment dynamics in the marine environment (Vol. 56). Amsterdam, Netherlands: Elsevier

Wu, W. (2016). Mixed cohesive and noncohesive sediment transport: a state of the art review. River sedimentation. CRC Press Taylor & Francis Group, London, 9-18.

Wu, Y., & Falconer, R., and Uncles, R. (1999). Modelling of water flows and cohesive sediment fluxes in the Humber Estuary, UK. Marine Pollution Bulletin, 37(3), 182-189.

Yangkhurung, J. R. (2018). Effect of canal operation on sedimentation and erosion in irrigation canals: a case study in Sunsari Morang irrigation system, Nepal. UNESCO-IHE,

Zac. (2012). Cohesive vs. Non-cohesive sediment experiments. In. People & Blogs: Daniel Duval.

Zhou, Z., & van der Wegen, M., & Jagers, B., and Coco, G. (2016). Modelling the role of self-weight consolidation on the morphodynamics of accretional mudflats. Environmental Modelling & Software, 76, 167-181.

elaud, G. and Baume, J. P. (2002). Maintaining equity in surface irrigation network affected by silt deposition. Journal of Irrigation and Drainage Engineering, 128(5), 316-325.

Bhutta, M. N., & Shahid, B. A., and Van Der Velde, E. J. (1996). Using a hydraulic model to prioritize secondary canal maintenance inputs: results from Punjab, Pakistan. Irrigation and Drainage Systems, 10(4), 377-392.

Caviedes-Voullième, D., & Morales-Hernández, M., & Juez, C., & Lacasta, A., and García-Navarro, P. (2017). Two-dimensional numerical simulation of bed-load transport of a finite-depth sediment layer: applications to channel flushing. Journal of Hydraulic Engineering, 143(9), 04017034.

Celik, I. and Rodi, W. (1988). Modeling suspended sediment transport in nonequilibrium situations. Journal of Hydraulic Engineering, 114(10), 1157-1191.

Chan, W., & Onyx, W., and Li, Y. (2006). Critical shear stress for deposition of cohesive sediments in Mai Po. Journal of Hydrodynamics, Ser. B, 18(3), 300-305.

Chow, T. V. (1959). Open-channel hydraulics (Vol. 1): McGraw-Hill New York.

Clemmens, A., & Bautista, E., & Wahlin, B., and Strand, R. (2005). Simulation of automatic canal control systems. Journal of Irrigation and Drainage Engineering, 131(4), 324-335.

Cole, P. and Miles, G. V. (1983). Two-dimensional model of mud transport. Journal of Hydraulic Engineering, 109(1), 1-12.

De Jong, J. (2005). Modelling the influence of vegetation on the morphodynamics of the river Allier: MSc thesis. Technology university of Delft,

Deltares. (2016). Delft3D-Flow user manual. Available online: https://oss.deltares.nl/documents/183920/185723/Delft3D-FLOW_User_Manual.pdf (accessed on 28-5-2016)

Department of Irrigation. (2003). Design Report Vol-I, Main Report and Appendices. Prepared by NEDECO for Sunsari Morang Irrigation Project Stage III (phase-I), Ministry of Water Resources, Nepal.

Depeweg, H. and Méndez, N. (2007). A New Approach to Sediment Transport in the Design and Operation of Irrigation Canals: UNESCO-IHE Lecture Note Series: CRC Press.

Depeweg, H. and Paudel, K. (2003). Sediment transport problems in Nepal evaluated by the SETRIC model. Irrigation and Drainage, 52(3), 247-260.

Depeweg, H., & Paudel, K. P., and Méndez, N. (2014). Sediment transport in irrigation canals: a new approach: CRC Press/ Balkema.

Devkota, L., & Crosato, A., and Giri, S. (2012). Effect of the barrage and embankments on flooding and channel avulsion case study Koshi River, Nepal. Rural Infrastructure 3 (3), 124-132.(2012).

DFID, M. M. (2006). Equity, Irrigation and Poverty, Guidelines for Sustainable Water Management, Final Report.

Dulovičová, R. and Velísková, Y. (2009). Aggradation of irrigation canal network in Žitný Ostrov, Southern Slovakia. Journal of Irrigation and Drainage Engineering, 136(6), 421-428.

Elias, E., & Walstra, D., & Roelvink, J., & Stive, M., and Klein, M. (2001). Hydrodynamic validation of Delft3D with field measurements at Egmond. Paper presented at the Coastal Engineering Conference.

Flokstra, C. (2006). Modelling of submerged vanes. Journal of Hydraulic Research, 44(5), 591-602.

Flokstra, C., & Jagers, H. R. A., & Wiersma, F. E., & Mosselman, E., and Jongeling, T. H. G. (2003). Numerical modelling of vanes and screens; development of vanes and screens in Delft3D-MOR. Retrieved from Delft University of Technology: http://resolver.tudelft.nl/uuid:304fa85d-bc41-46d3-948c-0e91bd91a6ab

Gebrehiwot, K. A., & Haile, A. M., & De Fraiture, C., & Chukalla, A. D., and Tesfa-alem, G. E. (2015). Optimizing flood and sediment management of spate irrigation in Aba'ala Plains. Water resources management, 29(3), 833-847.

Gismalla, Y. A. (2009). Sedimentation problems in the Blue Nile reservoirs and Gezira scheme: a review. Gezira Journal of Engineering and Applied Sciences, 4(2), 1-12.

Guan, W. B., & Wolanski, E., and Dong, L.-X. (1998). Cohesive sediment transport in the Jiaojiang River estuary, China. Estuarine, Coastal and Shelf Science, 46(6), 861-871.

Huang, & Jianchun, & Hilldale, R., and Greimann, B. (2006). Cohesive sediment transport. Erosion and sedimentation manual, 1-46.

Huang, J., & C., H. R., and P., G. B. (2008). Erosion And Sedimentation manual. Denver, Colorado U.S.: Department of the Interior Bureau of Reclamation.

Hung, M., & Hsieh, T., & Wu, C., and Yang, J. (2009). Two-dimensional nonequilibrium noncohesive and cohesive sediment transport model. Journal of Hydraulic Engineering, 135(5), 369-382.

Javernick, L., & Hicks, D., & Measures, R., & Caruso, B., and Brasington, J. (2016). Numerical modelling of braided rivers with structure-from-motion-derived terrain models. River Research and Applications, 32(5), 1071-1081.

Jian, H. U. (2008). Study on mathematical modeling for non-uniform sediment transport in an irrigation district along the lower reach of the Yellow River. Journal of China Institute of Water Resources and Hydropower Research, 1, 005.

Jinchi, H., & Zhaohui, W., and Qishun, Z. (1993). A study on sediment transport in an irrigation district. Paper presented at the 15th International Congress on Irrigation and Drainage, the Hague, the Netherlands 4-11 September.

Kemp, L. (2010). The evolution of sandbars along the Colorado River downstream of the Glen Canyon Dam: MSc thesis. Delft University of Technology, Delft.

Kondolf, G. M., & Gao, Y., & Annandale, G. W., & Morris, G. L., & Jiang, E., & Zhang, J., & Cao, Y., & Carling, P., & Fu, K., and Guo, Q. (2014). Sustainable sediment management in reservoirs and regulated rivers: Experiences from five continents. Earth's Future, 2(5), 256-280.

Krishnappan, B. G. (2000). Modelling cohesive sediment transport in rivers. IAHS Publication(International Association of Hydrological Sciences)(263), 269-276.

Krone, R. B. (1962). Flume studies of the transport of sediment in estuarial shoaling processes: University of California.

Lawrence, P., & Spark, and Counsell, C. (2001). Procedure for the selection and outline design of canal sediment control structures. HR Wallingford and Department for International Development (DFID).

Lawrence, P. A., Edmund. (1998). Deposition of fine sediments in irrigation canals. Irrigation and Drainage Systems, 12(4), 371-385. doi:10.1023/a:1006156926300

Lesser, G. R. (2009). An approach to medium-term coastal morphological modelling: UNESCO-IHE, PhD Thesis: CRC Press/ Balkema.

Lesser, G. R., & Roelvink, J. v., & Van Kester, J., and Stelling, G. (2004). Development and validation of a three-dimensional morphological model. Coastal engineering, 51(8-9), 883-915.

Li, L. (2010). A fundamental study of the Morphological Acceleration Factor. Civil Engineering and Geosciences. Retrieved from http://resolver.tudelft.nl/uuid:2780f537-402b-427a-9147-b8652279a83e, 23-8-2010

Liu, W. C., & Hsu, M. H., and Kuo, A. Y. (2002). Modelling of hydrodynamics and cohesive sediment transport in Tanshui River estuarine system, Taiwan. Marine Pollution Bulletin, 44(10), 1076-1088.

Lopes, J., & Dias, J., and Dekeyser, I. (2006). Numerical modelling of cohesive sediments transport in the Ria de Aveiro lagoon, Portugal. Journal of Hydrology, 319(1), 176-198.

Luijendijk, A. (2001). Validation, calibration and evaluation of Delft3D-FLOW model with ferry measurements.

Mendez, N. J. (1998). Sediment transport in irrigation canals: UNESCO-IHE, PhD Thesis: CRC Press/ Balkema.

Mishra, S. K. (2016). Draft Report On Main Irrigation Canal Operation Plan For Crops In Sitagunj Secondary Canal (S9) Irrigation System. IWRMP, DOI, Nepal.

Morianou, G. G., & Kourgialas, N. N., & Karatzas, G. P., and Nikolaidis, N. P. (2016). Hydraulic and sediment transport simulation of Koiliaris River using the MIKE 21C model. Procedia Engineering, 162, 463-470.

Munir, S. (2011). Role of Sediment Transport in Operation and Maintenance of Supply and Demand Based Irrigation Canals: Application to Machai Maira Branch Canals: UNESCO-IHE, PhD Thesis: CRC Press/ Balkema.

Nawazbhutta, M., & Shahid, B. A., and Van Der Velde, E. J. (1996). Using a hydraulic model to prioritize secondary canal maintenance inputs: results from Punjab, Pakistan. Irrigation and Drainage Systems, 10(4), 377-392.

Nicholson, J. and O'Connor, B. A. (1986). Cohesive sediment transport model. Journal of Hydraulic Engineering, 112(7), 621-640.

Nippon Koei. (1995). Project Operation Plan. Sunsari Morang Irrigation Project, Ministry of Water Resources, Nepal.

Osman, & S., I., & Schultz, B., & Osman, A., and Suryadi, F. (2017). Effects of different operation scenarios on sedimentation in irrigation canals of the Gezira Scheme, Sudan. Irrigation and Drainage, 66(1), 82-89.

Osman, I. (2015). Impact of improved operation and maintenance on cohesive sediment transport in Gezira Scheme, Sudan: UNESCO-IHE, PhD Thesis: CRC Press/ Balkema.

Osman, I. S., & Schultz, B., & Osman, A., and Suryadi, F. (2016). Simulation of Fine Sediment Transport in Irrigation Canals of the Gezira Scheme with the Numerical Model FSEDT. Journal of Irrigation and Drainage Engineering, 142(11).

Osman, I. S. E. (2012). Sediment and Water Management of Large Irrigation Systems, Case Study: Gezira Scheme, Sudan. (PhD proposal), UNESCO-IHE institute, Delft.

Parsapour-Moghaddam, P. and Rennie, C. D. (2017). Hydrostatic versus nonhydrostatic hydrodynamic modelling of secondary flow in a tortuously meandering river: Application of Delft3D. River research and applications, 33(9), 1400-1410.

Partheniades, E. (1965). Erosion and deposition of cohesive soils. Journal of the Hydraulics Division, 91(1), 105-139.

Partheniades, E. (1986). A fundamental framework for cohesive sediment dynamics. In *Estuarine cohesive sediment dynamics* (pp. 219-250): Springer.

Partheniades, E. (2009). Cohesive Sediments in Open Channels: Erosion, Transport and Deposition. United Kingdom: Butterworth-Heinemann.

Paudel, K. P. (2010). Role of Sediment in the Design and Management of Irrigation Canals: Sunsari Morang Irrigation Scheme, Nepal: UNESCO-IHE, PhD Thesis: CRC Press/Balkema.

Paudel, K. P., & Depeweg, H., and Méndez, N. (2014). Sediment transport in irrigation canals: a new approach: CRC Press.

Ravisangar, V., & Sturm, T., and Amirtharajah, A. (2005). Influence of sediment structure on erosional strength and density of kaolinite sediment beds. Journal of Hydraulic Engineering, 131(5), 356-365.

Renault, D., & Facon, T., and Wahaj, R. (2007). Modernizing Irrigation Management: The MASSCOTE Approach--Mapping System and Services for Canal Operation Techniques (Vol. 63): Food & Agriculture Org.

Renault, D. and Wahaj, R. (2006). MASSCOT: a methodology to modernize irrigation services and operation in canal systems. Applications to two systems in Nepal Terai: Sunsari Morang Irrigation System and Narayani Irrigation System. FAO. Rome, 1-42.

Roelvink, D., & Boutmy, A., and Stam, J.-M. (1998). A simple method to predict long-term morphological changes. Coastal Engineering Proceedings, 1(26).

Roelvink, J. and Van Banning, G. (1995). Design and development of DELFT3D and application to coastal morphodynamics. Oceanographic Literature Review, 11(42), 925.

Sanford, L. P. and Halka, J. P. (1993). Assessing the paradigm of mutually exclusive erosion and deposition of mud, with examples from upper Chesapeake Bay. Marine Geology, 114(1-2), 37-57.

Schaffner, J. (2008). Numerical investigations on the function of flush waves in a reservoir sewer. Technische Universität, Darmstadt, Germany.

Schultz, B. (2002). Land and Water Development. Delft, the Netherlands: IHE.

Schultz, B. and De Wrachien, D. (2002). Irrigation and drainage systems research and development in the 21st century. Irrigation and Drainage: The journal of the International Commission on Irrigation and Drainage, 51(4), 311-327.

Sherpa, K. (2005). Use of Sediment Transport Model SETRIC in an Irrigation Canal. Unesco-IHE, Delft, Netherlands.

Simons, D. and Fuat, S. (1992). Sediment Trasport Technology. Colorado, USA.

Sutama, N. (2010). Mathematical modelling of sediment transport and its improvement in Bekasi Irrigation System, WestJava, Indonesia. MSc thesis Unesco-IHE.

Teisson, C. M. O., P. Le Hir, C. Kranenburg, L. Hamm,. (1993). Cohesive sediment transport processes,. Coastal Engineering, (http://www.sciencedirect.com/science/article/pii/037838399390048D), Volume 21(Issues 1–3), 34 Pages. doi:https://doi.org/10.1016/0378-3839(93)90048-D.

Thanh, N. V., & Chung, D. H., and Nghien, T. D. (2014). Prediction of the local scour at the bridge square pier using a 3D numerical model. Open Journal of Applied Sciences, 4(02), 34.

Theol, & A, S., & Jagers, B., & Suryadi, F., and de Fraiture, C. (2019). The use of Delft3D for Irrigation Systems Simulations. Irrigation and Drainage, 68(2), 318-331. doi:10.1002/ird.2311

Theol, S., & Bert Jagers, & F. Suryadi, and Charlotte De Fraiture. (2019a). The role of operation in reducing problems with cohesive and non-cohesive sediments in irrigation canals. MDPI/Water Journal.

Theol, S., & Bert Jagers, & F. Suryadi, and Charlotte De Fraiture. (2019b). The use of Delft3D for Irrigation Systems Simulations. Irrigation and Drainage, 68(2), 318-331.

Theol, S., & Bert Jagers, & F. Suryadi, and Charlotte De Fraiture. (forthcoming). The use of 2D/3D models to show the differences between cohesive and non-cohesive sediments in irrigation systems. Submitted to the American Journal of Irrigation and Drainage Engineering (ASCE).

Van der Wegen, M., & Jaffe, B., and Roelvink, J. (2011). Process-based, morphodynamic hindcast of decadal deposition patterns in San Pablo Bay, California, 1856–1887. Journal of Geophysical Research: Earth Surface, 116(F2).

Van Leussen, W. (1994). Estuarine macroflocs and their role in fine-grained sediment transport: Ministry of Transport, Public Works and Water Management, National Institute for Coastal and Marine Management (RIKZ).

Van Rijn, L. C. (1993). Principles of sediment transport in rivers, estuaries and coastal seas (Vol. 1006): Aqua publications Amsterdam.

Van Rijn, L. C., & Van Rossum, H., and Termes, P. (1990). Field verification of 2-D and 3-D suspended-sediment models. Journal of Hydraulic Engineering, 116(10), 1270-1288.

Villaret, C., & Hervouet, J.-M., & Kopmann, R., & Merkel, U., and Davies, A. G. (2013). Morphodynamic modeling using the Telemac finite-element system. Computers & Geosciences, 53, 105-113.

Winterwerp, J. C. and Van Kesteren, W. G. (2004). Introduction to the physics of cohesive sediment dynamics in the marine environment (Vol. 56). Amsterdam, Netherlands: Elsevier

Wu, W. (2016). Mixed cohesive and noncohesive sediment transport: a state of the art review. River sedimentation. CRC Press Taylor & Francis Group, London, 9-18.

Wu, Y., & Falconer, R., and Uncles, R. (1999). Modelling of water flows and cohesive sediment fluxes in the Humber Estuary, UK. Marine Pollution Bulletin, 37(3), 182-189.

Yangkhurung, J. R. (2018). Effect of canal operation on sedimentation and erosion in irrigation canals: a case study in Sunsari Morang irrigation system, Nepal. UNESCO-IHE,

Zac. (2012). Cohesive vs. Non-cohesive sediment experiments. In. People & Blogs: Daniel Duval.

Zhou, Z., & van der Wegen, M., & Jagers, B., and Coco, G. (2016). Modelling the role of self-weight consolidation on the morphodynamics of accretional mudflats. Environmental Modelling & Software, 76, 167-181.

LIST OF ACRONYMS

1D	One Dimensional
2D	Two Dimensional
3D	Three Dimensional
A	Area (m^2)
$A^{(l)}$	Rouse number
a	Van rijn's reference height (m)
B	Bottom width (m)
C_a^l	Reference concentration of sediment fraction (l) $[kg/m^3]$;
C^l_{kmx}	Average concentration of the kmx cell of sediment fraction (l) $[kg/m^3]$;
C	Total sediment concentration at distance x (ppm)
Ce	Total sediment concentration in equilibrium condition (ppm)
C_0	Total sediment concentration at distance x=0 (ppm)
c_e	Concentration of suspended load in equilibrium condition (ppm)
c	Concentration of suspended load at distance x (ppm)
C_b	Bed load concentration (m^3/m^3)
C'	Chezy coefficient related to grains $(m^{1/2}/s)$
ca	Reference concentration (m^3/m^3)
d_{50}	Median diameter of material (m)
D^l	Deposition flux $(kg\ m^{-2}s^{-1})$
E^l	Erosion flux $(kg\ m^{-2}s^{-1})$
f	User define proportionality factor;[-]
g	Acceleration due to gravity (m/s^2)
h	Water depth (m)
k_s	Equivalent roughness height (m)
κ	Constant of von Karman (dimensionless)
k	Empirical constant (-)
M^l	User-defined erosion parameter $(kg\ m^{-2}s^{-1})$
n	an exponential (-)
n	Manning's roughness coefficient $(s/m^{1/3})$
N	Exponent constant of von Karman (dimensionless)
Q	Discharge (m^3/s)

Q_s	Sediment discharge (m³/s)
R	Hydraulic radius (m)
s	Relative density ($\rho s/\rho$) (dimensionless)
S_b	Thickness of bed load layer (m)
S_o	Longitudinal slope for canal (-)
S_s	Side slope for trapezoidal canals (-)
$S(\tau_{cw}, \tau_{cr,d}^l)$	Deposition step function,
T	Bed shear parameter (dimensionless)
t	Time (s)
$u*_{cr}$	Critical bed shear velocity (m/s)
u	Water flow velocity in x (horizontal) direction (m/s)
V	Mean velocity (m/s)
w_s^l	Fall velocity (hindered) [m/s]
x	Horizontal coordinates (m)
	Vertical coordinates (m)
z	Bottom level above datum (m)
z	roughness length, which can be obtained from Nikuradsa (K_s)
Z_o,	Depth down to the bed from a reference height (positive down)[m]
Z_b	Thickness of the bed layer[m]
ΔZ_b	Difference in elevation between the center of the kmx cell, Van
Δz	Rijn's reference height
Δt	Time step / time interval (s)
$\propto_1^l$	Correction factor [-]
$\propto_2^l$	Correction factor for sediment concentration [-]
ε_s^l	Sediment diffusion coefficient evaluated at the bottom of the kmx cell of sediment fraction (l) [-]
θ	Dimensionless mobility parameter (-)
φ	Dimensionless transport parameter (-)
$\tau*_o$	Critical dimensionless shear stress (N/m²)
τ	Local bed shear stress (N/m²)
τ_{cr}	Critical shear stress (N/m²)
ρ	Density (kg/m³)

β	Rratio of sediment and fluid mixing coefficient (-)
εx	Sediment mixing coefficient in x direction (m^2/s)
εz	Sediment mixing coefficient in z direction (m^2/s)
τ_{cw}	Maximum bed shear stress [N/m^2]
$\tau_{cr,e}^{l}$	User-defined critical erosion shear stress [N/m^2]
$\tau_{cr,d}^{l}$	User-defined critical deposition shear stress [N/m^2]

LIST OF TABLES

LIST OF FIGURES

ABOUT THE AUTHOR

Shaimaa Abd Al-Amear Theol was born in Baghdad, Iraq. She graduated from Baghdad University/ college of Engineering from Irrigation and Drainage Department in 2003. From 2004, Shaimaa started to work in Iraqi Ministry of Water Resources at the General Authority for Groundwater as an engineer. In 2009 Shaimaa started study as Master student in Water Science and Engineering Department in IHE-Delft (was named UNESCO-IHE), Delft, the Netherlands and she got her MSc diploma in 2011. From August 2013 Shaimaa became a Ph.D. fellow in the Water Science and Engineering Department/ Land and Water Development for Food Security in IHE-Delft, and Wageningen University & research in the Netherlands. Her study is fully supported by the Iraqi Ministry of Higher Education and Scientific Research. She has a background in engineering. Her primary research interests are about the cohesive and non-cohesive impacts on the irrigation systems, and about the 2D/3D modelling in irrigation systems.

Journals publications

- Theol, & A, S., & Jagers, B., & Suryadi, F., and de Fraiture, C. (2019). The use of Delft3D for Irrigation Systems Simulations. Irrigation and Drainage, 68(2), 318-331. doi:10.1002/ird.2311.

- Theol, Shaimaa, & Bert Jagers, & F. Suryadi, and Charlotte De Fraiture. The role of gate operation in reducing problems with cohesive and non-cohesive sediments in irrigation canals. *Water* **2019**, *11*, 2572; doi: 10.3390/w11122572.

- Theol, Shaimaa, Bert Jagers, F.X. Suryadi, Charlotte de Fraiture. The use of 2D/3D models to show the differences between cohesive and non-cohesive sediments in irrigation systems. Submitted (In second review) to the American Journal of Irrigation and Drainage Engineering (ASCE).

- Theol, Shaimaa, & Bert Jagers, & J. Rai & F. Suryadi, and Charlotte De Fraiture. What is the effect of gate selection on the non-cohesive sedimentation in an irrigation schemes? Submitted to the Water Journal (MDPI).

Conference proceedings

- Theol, Shaimaa, & Charlotte de Fraiture, F. Suryadi (2016) Understanding cohesive sediments behaviour in irrigation canals using Delft3D model simulation. Paper presented at the Second World Irrigation Forum in Chiang Mai, Thailand, Conference Abstract.

- Theol, Shaimaa, & Bert Jagers, & Charlotte de Fraiture and F. Suryadi (2017). Proving the possibility of using Delft3D in irrigation systems simulations. Paper presented at the EGU General Assembly Conference Abstracts.

- Theol, Shaimaa, & Bert Jagers, & Charlotte de Fraiture and F. Suryadi (2018). Simulating sediment transport in irrigation systems using Delft3D. Paper presented at DSD conference in Deltares, Delft, Netherlands Conference Abstracts.

**Netherlands Research School for the
Socio-Economic and Natural Sciences of the Environment**

D I P L O M A

For specialised PhD training

The Netherlands Research School for the
Socio-Economic and Natural Sciences of the Environment
(SENSE) declares that

Shaimaa Abd Al-Amear Theol

born on 28 May 1977 in Baghdad, Iraq

has successfully fulfilled all requirements of the
Educational Programme of SENSE.

Delft, 19 February 2020

The Chairman of the SENSE board

Prof. dr. Martin Wassen

the SENSE Director of Education

Dr. Ad van Dommelen

The SENSE Research School declares that Shaimaa Abd Al-Amear Theol has successfully
fulfilled all requirements of the Educational PhD Programme of SENSE with a
work load of 34.9 EC, including the following activities:

SENSE PhD Courses

o Environmental research in context (2014)
o Research in context activity: 'Co-organizing UNESCO-IHE PhD symposium and
  Proceedings Booklet (Delft, 28-29 September 2015)'
o Sense writing week (2015)
o Summer school Communicating water; bridging the gap between science and society
  (2016)
o Grasping Sustainability (2017)

Other PhD and Advanced MSc Courses

o Conveyance and irrigation structures, IHE Delft (2016)
o High performance computing helping to solve water related problems, IHE Delft (2017)
o Basic and advanced course on "Morphological modelling using Delft3D model, IHE Delft
  (2014)
o
External training at a foreign research institute

o Delft Software days 2016 and 2017, Deltares, The Netherlands

Management and Didactic Skills Training

o Supervising MSc student with thesis entitled ' Effect of Canal Operation on
  Sedimentation and Erosion in Irrigation Canals (A Case Study in Sunsari Morang
  Irrigation System, Nepal)' (2018)
o Teaching in the MSc course ' Module 7: Conveyance and Irrigation Structures' (2018)

Oral Presentations

o *Effects of cohesive sedimentation in the irrigation systems.* IHE PhD symposium, 29
  September- 3 October 2014, Delft, the Netherlands
o *Understanding the difference in behaviour between the cohesive and non-cohesive
  sediments.* IHE PhD symposium, 28-29 September 2015, Delft, The Netherlands
o *Simulating sediment transport in irrigation systems using Delft3D.* Deltares Delft
  Software Days, 12 November 2018, Delft, The Netherlands

SENSE Coordinator PhD Education

Dr. Peter Vermeulen

Printed and bound by CPI Group (UK) Ltd, Croydon, CR0 4YY

24/10/2024

01778309-0002